# STATUS AND TRENDS IN SPENT FUEL AND RADIOACTIVE WASTE MANAGEMENT

The following States are Members of the International Atomic Energy Agency:

| | | |
|---|---|---|
| AFGHANISTAN | GEORGIA | PAKISTAN |
| ALBANIA | GERMANY | PALAU |
| ALGERIA | GHANA | PANAMA |
| ANGOLA | GREECE | PAPUA NEW GUINEA |
| ANTIGUA AND BARBUDA | GRENADA | PARAGUAY |
| ARGENTINA | GUATEMALA | PERU |
| ARMENIA | GUINEA | PHILIPPINES |
| AUSTRALIA | GUYANA | POLAND |
| AUSTRIA | HAITI | PORTUGAL |
| AZERBAIJAN | HOLY SEE | QATAR |
| BAHAMAS, THE | HONDURAS | REPUBLIC OF MOLDOVA |
| BAHRAIN | HUNGARY | ROMANIA |
| BANGLADESH | ICELAND | RUSSIAN FEDERATION |
| BARBADOS | INDIA | RWANDA |
| BELARUS | INDONESIA | SAINT KITTS AND NEVIS |
| BELGIUM | IRAN, ISLAMIC REPUBLIC OF | SAINT LUCIA |
| BELIZE | IRAQ | SAINT VINCENT AND |
| BENIN | IRELAND | THE GRENADINES |
| BOLIVIA, PLURINATIONAL | ISRAEL | SAMOA |
| STATE OF | ITALY | SAN MARINO |
| BOSNIA AND HERZEGOVINA | JAMAICA | SAUDI ARABIA |
| BOTSWANA | JAPAN | SENEGAL |
| BRAZIL | JORDAN | SERBIA |
| BRUNEI DARUSSALAM | KAZAKHSTAN | SEYCHELLES |
| BULGARIA | KENYA | SIERRA LEONE |
| BURKINA FASO | KOREA, REPUBLIC OF | SINGAPORE |
| BURUNDI | KUWAIT | SLOVAKIA |
| CABO VERDE | KYRGYZSTAN | SLOVENIA |
| CAMBODIA | LAO PEOPLE'S DEMOCRATIC | SOMALIA |
| CAMEROON | REPUBLIC | SOUTH AFRICA |
| CANADA | LATVIA | SPAIN |
| CENTRAL AFRICAN | LEBANON | SRI LANKA |
| REPUBLIC | LESOTHO | SUDAN |
| CHAD | LIBERIA | SWEDEN |
| CHILE | LIBYA | SWITZERLAND |
| CHINA | LIECHTENSTEIN | SYRIAN ARAB REPUBLIC |
| COLOMBIA | LITHUANIA | TAJIKISTAN |
| COMOROS | LUXEMBOURG | THAILAND |
| CONGO | MADAGASCAR | TOGO |
| COOK ISLANDS | MALAWI | TONGA |
| COSTA RICA | MALAYSIA | TRINIDAD AND TOBAGO |
| CÔTE D'IVOIRE | MALI | TUNISIA |
| CROATIA | MALTA | TÜRKİYE |
| CUBA | MARSHALL ISLANDS | TURKMENISTAN |
| CYPRUS | MAURITANIA | UGANDA |
| CZECH REPUBLIC | MAURITIUS | UKRAINE |
| DEMOCRATIC REPUBLIC | MEXICO | UNITED ARAB EMIRATES |
| OF THE CONGO | MONACO | UNITED KINGDOM OF |
| DENMARK | MONGOLIA | GREAT BRITAIN AND |
| DJIBOUTI | MONTENEGRO | NORTHERN IRELAND |
| DOMINICA | MOROCCO | UNITED REPUBLIC OF TANZANIA |
| DOMINICAN REPUBLIC | MOZAMBIQUE | UNITED STATES OF AMERICA |
| ECUADOR | MYANMAR | URUGUAY |
| EGYPT | NAMIBIA | UZBEKISTAN |
| EL SALVADOR | NEPAL | VANUATU |
| ERITREA | NETHERLANDS, | VENEZUELA, BOLIVARIAN |
| ESTONIA | KINGDOM OF THE | REPUBLIC OF |
| ESWATINI | NEW ZEALAND | VIET NAM |
| ETHIOPIA | NICARAGUA | YEMEN |
| FIJI | NIGER | ZAMBIA |
| FINLAND | NIGERIA | ZIMBABWE |
| FRANCE | NORTH MACEDONIA | |
| GABON | NORWAY | |
| GAMBIA, THE | OMAN | |

IAEA Nuclear Energy Series No. NW-T-1.14 (Rev. 2)

# STATUS AND TRENDS IN SPENT FUEL AND RADIOACTIVE WASTE MANAGEMENT

INTERNATIONAL ATOMIC ENERGY AGENCY

VIENNA, 2025

# COPYRIGHT NOTICE




**IAEA Library Cataloguing in Publication Data**

Names: International Atomic Energy Agency.
Title: Status and trends in spent fuel and radioactive waste management / International Atomic Energy Agency.
Description: Vienna : International Atomic Energy Agency, 2025. | Series: IAEA nuclear energy series, ISSN 1995-7807 ; no. NW-T-1.14 (rev. 2) | Includes bibliographical references.
Identifiers: IAEAL 25-01797 | ISBN 978-92-0-104725-0 (paperback : alk. paper) | ISBN 978-92-0-104925-4 (pdf) | ISBN 978-92-0-104825-7 (epub)
Subjects: LCSH: Radioactive wastes — Management. | Spent reactor fuels. | Radioactive waste disposal.
Classification: UDC 621.039.7 | STI/PUB/2109

# FOREWORD

The IAEA's statutory role is to "seek to accelerate and enlarge the contribution of atomic energy to peace, health and prosperity throughout the world". Among other functions, the IAEA is authorized to "foster the exchange of scientific and technical information on peaceful uses of atomic energy". One way this is achieved is through a range of technical publications, including the IAEA Nuclear Energy Series.

The IAEA Nuclear Energy Series comprises publications designed to further the use of nuclear technologies in support of sustainable development, to advance nuclear science and technology, catalyse innovation and build capacity to support both the existing and expanded use of nuclear power and nuclear science applications. These publications address policy, technological and management aspects of the definition and implementation of activities involving the peaceful use of nuclear technology.

The IAEA Safety Standards establish fundamental principles, requirements and recommendations to ensure nuclear safety and serve as a global reference for protecting people and the environment from the harmful effects of ionizing radiation.

When IAEA Nuclear Energy Series publications address safety, the IAEA Safety Standards are referred to as the prevailing boundary conditions for the application of nuclear technology.

Building on this foundation, the IAEA works with Member States to ensure the safe, secure and sustainable management of spent fuel and radioactive waste. Radioactive material is used worldwide in medicine, industry, research and power generation. These applications inevitably produce radioactive waste, which must be managed safely and, where appropriate, disposed of in a manner that ensures long term protection of people and the environment.

Preparation of Status and Trends in Spent Fuel and Radioactive Waste Management is a collaborative project between the IAEA, the European Commission and the OECD Nuclear Energy Agency, with the participation of the World Nuclear Association. The project brings together global data and expert analysis to provide a coherent, authoritative picture of how Member States manage their spent fuel and radioactive waste. The first edition, published in January 2018, covered the situation up to the end of 2013. The second, released in 2022, reflected developments up to the end of 2016. This third edition covers the period to the end of 2019.

This publication presents global inventories of spent fuel and radioactive waste, describes current management arrangements, and outlines future plans for their ultimate disposal where appropriate. It addresses institutional, organizational and technical aspects, including legal and regulatory systems, allocation of responsibilities, and strategies for managing all types of spent fuel and radioactive waste from generation through conditioning and storage to disposal.

The IAEA acknowledges with appreciation the contributions of Member State experts, collaborating organizations and all those who supported the preparation of this edition. Special thanks are extended to E. Garcia Neri, who chaired the work for this cycle. The IAEA officer responsible for this publication was M. Lust of the Division of Nuclear Fuel Cycle and Waste Technology.

# CONTENTS

# SUMMARY


Radioactive material is used to treat cancer, monitor the quality of industrial products and generate electricity (among other beneficial uses). In common with all processes, some waste arises from these applications. The waste comprises various forms and materials, with different radioactivity levels and half-lives. Radioactive waste needs to be handled safely and eventually disposed of in a safe manner. Acceptable disposal routes depend on the level of radioactivity and established preferences and practices in different countries. Some waste contains such low levels of radioactivity that it can be released from regulatory control and disposed of as non-radioactive waste. However, for radioactive waste that presents a long-term risk to people and the environment, its end point is placement in an appropriate package and disposal in a suitably engineered, multibarrier facility.

Status and Trends in Spent Fuel and Radioactive Waste Management is a collaborative project between the IAEA, the European Commission and the Organisation for Economic Co-operation and Development (OECD) Nuclear Energy Agency, with the participation of nuclear industry organization the World Nuclear Association, that aims to consolidate and complement the information gathered from different initiatives around the world. The objective of the Status and Trends in Spent Fuel and Radioactive Waste Management series is to be the authoritative publication that systematically and periodically summarizes the global status and trends of programmes and inventories for spent fuel and radioactive waste management. The first in the series was published in January 2018 [1] and covered the situation up to the end of December 2013. The second report was published in 2022 [2] and covered the situation up to the end of December 2016. This current report is the third edition and covers the situation up to the end of December 2019.

This publication provides an overview of current global inventories of spent fuel and radioactive waste, current arrangements for their management and future plans for their ultimate disposal where appropriate. Spent fuel is generated only by Member States operating nuclear power plants or non-power reactors, whereas radioactive waste is generated in all States producing or using radioactive material in, for example, medicine, industry and research and the nuclear fuel cycle. It is the intention to update this publication at regular intervals, following the reporting schedule for the Joint Convention on the Safety of Spent Fuel Management and on the Safety of Radioactive Waste Management (Joint Convention) [3].

Institutional, organizational and technical aspects of spent fuel and radioactive waste management are explored, including legal and regulatory systems; organization of waste management activities and associated responsibilities; and strategies and plans for ongoing management of different types of spent fuel and radioactive waste, from its generation through conditioning and storage to disposal. This publication compiles the quantities of spent fuel and radioactive waste that currently exist and explores forecasts for the coming decades. Significant trends and the corresponding challenges in the management of spent fuel and radioactive waste are also discussed.

Inventory estimates of spent fuel and radioactive waste in the world are based on information in the Spent Fuel and Radioactive Waste Information System[1] provided by 19 participating Member States. Data are supplemented by published reports to the Joint Convention. For most cases, the information provided corresponds to the end of December 2019; the data are based on information from States accounting for almost 92% of all nuclear power reactors in the world. On this basis, there is an estimated 301 000 tonnes of heavy metal (t HM) of spent fuel in storage worldwide. The current total global inventory of solid radioactive waste is approximately 32 million $m^3$, of which 26.6 million $m^3$ (83% of the total) has been disposed of permanently and a further 5.6 million $m^3$ (17%) is in storage awaiting final disposal. More than 92% of the volume of solid waste is classified as being very low or low level waste, with most of the remainder being intermediate level waste. In terms of total radioactivity, the situation is fully reversed, with approximately 95% of the radioactivity being associated with intermediate and high level waste.

If naturally occurring radioactive material (NORM) is classified as radioactive waste, depending on the national waste management concept, this is usually considered to be very low level waste (VLLW)


---

[1] See sris.iaea.org

or low level waste (LLW). NORM waste is not specifically discussed in this publication, although some countries have reported NORM waste in the Spent Fuel and Radioactive Waste Information System.

It is evident that significant progress has been made globally in formulating national policies and strategies and in implementing legal and regulatory systems that define responsibilities for the ongoing safe management of spent fuel and radioactive waste. Most States expect to dispose of their waste in facilities located on their territories, with the main focus of international cooperation being on technology development. Disposal facilities for VLLW and LLW are already in operation in several countries. However, in many others, particularly those with small volumes of radioactive waste, disposal options still have to be developed. The most important remaining challenge is the development, public acceptance and long term funding of disposal facilities for high level waste and spent nuclear fuel considered as waste. Significant progress has been made in a few countries. In 2022 the construction licence for the deep geological repository (DGR) was approved in Sweden, the operational licence was submitted for a deep geological disposal facility in Finland and France recognized a deep geological disposal facility as the final disposal solution for high level and intermediate level long lived radioactive waste.

# 1. INTRODUCTION

## 1.1. BACKGROUND

Spent fuel and radioactive waste are by-products of the operation of nuclear reactors and related fuel cycle activities, and other uses of radioactive material in medicine, industry and research. Currently two strategies are employed for managing spent fuel from power reactors: either it is considered to be waste or it is considered to be an asset. In the latter case, additional treatment is necessary to recover uranium and plutonium, generating high level waste as a by-product. The radioactive waste comprises various forms and materials, with different radioactivity levels and half-lives.

According to IAEA guidance provided in IAEA Safety Standards Series No. GSG-1, Classification of Radioactive Waste [4], the management and disposal options for radioactive waste are dependent upon its classification: high level waste (HLW), intermediate level waste (ILW), low level waste (LLW) or very low level waste (VLLW). The final disposal of the waste may range from geological disposal for HLW to near surface trench disposal for VLLW. The activity level and the nature of radionuclides in the waste, as well as waste properties, determine the conditioning needs of the waste before disposal. There are also possibilities, as the case may be, release from regulatory control.

Status and Trends in Spent Fuel and Radioactive Waste Management (hereafter referred to as the Status and Trends project) is a collaborative project between the IAEA, the European Commission and the Organisation for Economic Co-operation and Development (OECD) Nuclear Energy Agency (OECD/NEA), with the participation of the nuclear industry organization the World Nuclear Association (WNA), that aims to consolidate and complement the information gathered from different initiatives around the world. This publication series aims to provide authoritative publications that systematically and periodically summarize the global status and trends of programmes and inventories for spent fuel and radioactive waste management. The first publication in the series was published in January 2018 [1] and covered the situation up to the end of December 2013. The second publication was published in January 2022 [2], covering the situation up to the end of December 2016. This current publication is the second revision and covers the situation up to the end of December 2019. The analysis presented is based on information as of December 2019 to be consistent with the information in reports submitted by 2020 under the framework of the Joint Convention on the Safety of Spent Fuel Management and on the Safety of Radioactive Waste Management (hereafter referred to as the Joint Convention) [3] and those provided to the European Commission in 2021 in accordance with Council Directive 2011/70/Euratom of 19 July 2011, establishing a Community framework for the responsible and safe management of spent fuel and radioactive waste (hereafter referred to as the European Atomic Energy Community (Euratom) Waste Directive) [5]. The basic information in this publication has been collected through submission to Spent Fuel and Radioactive Waste Information System (SRIS) and has been complemented with openly available Joint Convention National Reports. Approximately 90% of States with operating nuclear power plants submitted their data to SRIS or their Joint Convention National Reports are openly available, representing almost 92% of all nuclear power reactors in the world.

Certain publications cover the subject on a partial or regional basis. The European Commission has published such information every three years since 1992 (see Ref. [6]). The OECD/NEA publishes profiles and reports about radioactive waste management programmes in member countries[1], as well as an annual Nuclear Energy Data report on nuclear power status in NEA member countries and the OECD area [7].

This publication goes further by providing an extensive overview of the management of spent fuel and radioactive waste worldwide and of the quantities involved. The volumes of different types of waste give an indication of the magnitude of the work needed for managing and disposing. A large volume does not, however, necessarily correspond to a large risk or environmental impact. In particular, when

---

[1] See www.oecd-nea.org/rwm/profiles

determining the potential impact of different materials and waste classes, the radioactivity content needs to be considered. Chemical and other hazardous properties of the waste also need to be considered.

## 1.2. OBJECTIVE

The purpose of this publication is to provide a global overview of the status of spent fuel and radioactive waste management programmes, inventories, current practices, technologies and trends. It provides overviews of national arrangements for the management of spent fuel and radioactive waste, and of current waste and spent fuel inventories. Achievements, challenges and trends in the management of spent fuel and radioactive waste are also addressed. The data reported are fully dependent on the input from the Member States and the assumptions made to transform these data into the waste classes defined in GSG-1 [4].

The anticipated audience for the Status and Trends publications includes national policy and decision makers and their support staff, as well as professionals in the nuclear and other scientific fields who wish to obtain an overview of how the spent fuel and radioactive waste are handled in different countries. Although the publication is not directly written for a general audience, much of the information contained herein could be of interest to the public, including the media, researchers, educators and students.

Guidance and recommendations provided here in relation to identified good practices represent expert opinion but are not made on the basis of a consensus of all Member States.

## 1.3. SCOPE

This publication addresses the institutional, legal and regulatory frameworks for the management of spent fuel and radioactive waste; spent fuel and radioactive waste management programmes, current practices and technologies; and spent fuel and radioactive waste inventories and forecasts. In addition, this publication provides an analysis of the trends and the achievements made in the frame of spent fuel management and radioactive waste management, together with a view on the challenges that are yet to be overcome.

The publication includes all material that a Member State has declared as being radioactive waste, along with spent nuclear fuel (whether the spent fuel has been declared to be waste or not). The collection and compilation of information on spent fuel and radioactive waste reflects the different strategies for spent fuel management (open cycle, closed cycle or awaiting decision) pursued by various countries and involves several challenges, such as the use of different waste classification schemes in different countries, different statuses of waste conditioning and different stages of development of waste management systems.

The following types of radioactive materials are specifically excluded:

— Exempt waste that meets the criteria for clearance, exemption or exclusion from regulatory control for radiation protection purposes (e.g. as per GSG-1 [4]);
— Very short lived waste (since this is typically held for decay and then released);
— Authorized effluent releases;
— Radioactive materials and products that have not been declared radioactive waste;
— Radioactive waste from extractive industries (e.g. uranium mining, NORM) if they have not been included in the relevant national inventory;
— Waste from military/defence activities if they have not been included in the relevant national inventory;
— Contaminated sites and related buildings, when these are not included in the relevant national inventory.

Other sources of information, in addition to the SRIS, include the following:

— Openly available National Reports to the Joint Convention, which are produced every three years in advance of the Review Meetings of the Contracting Parties. The IAEA provides the secretariat for the Joint Convention and the latest Review Meeting was held in 2022;
— Openly available National Reports to the European Commission in accordance with the Euratom Waste Directive. The latest reporting to the European Commission according to this Directive was due in August 2021.

## 1.4. STRUCTURE

This publication provides a global overview of current spent fuel and radioactive waste management and provides a compilation of policies, strategies and spent fuel and radioactive waste quantities on a regional and global scale. This is based on information in the SRIS supplied by the Member State or on information provided by the State in its report to the Joint Convention. Section 2 presents the relevant international legal instruments to improve the safety of spent fuel and radioactive waste management. Section 3 outlines the sources of spent fuel and radioactive waste. Sections 4 and 5 explore the frameworks for managing these and give a summary of current strategies, practices and technologies. Section 6 describes the inventories and presents future forecasts. Sections 7 and 8 provide analysis and achievements, as well as trends and challenges, and Section 9 concludes.

# 2. INTERNATIONAL LEGAL INSTRUMENTS AND SUPPORTING MATERIALS

There are several international instruments to ensure and improve the safe management of spent fuel and radioactive waste, and to protect people and the environment from the potential for negative effects of ionizing radiation. These instruments include, inter alia, international conventions, standards and peer reviews. All of them are supported by publications produced by the academic community, the organizations involved in the management of spent fuel and radioactive waste, different agreements between organizations and countries, etc. This chapter gives a short overview of the main legal instruments and supporting materials.

## 2.1. JOINT CONVENTION ON THE SAFETY OF SPENT FUEL MANAGEMENT AND ON THE SAFETY OF RADIOACTIVE WASTE MANAGEMENT

The Joint Convention on the Safety of Spent Fuel Management and on the Safety of Radioactive Waste Management (Joint Convention) [3], which entered into force in 2001, highlights the importance given to spent fuel and radioactive waste management. The Joint Convention applies to spent fuel and radioactive waste created through the processes of civilian nuclear programmes, while spent fuel and other radioactive waste resulting from military or defence programmes usually fall under the Convention when they are transferred permanently to and managed within exclusively civilian programmes. As of October 2022, there were 88 Contracting Parties to the Joint Convention. The Contracting Parties meet every three years to discuss the National Reports, which are subject to a peer review process.[2] There have already been seven review meetings, the last of which was held from 27 June–8 July 2022 [8]. The next review meeting is scheduled in 2025.

## 2.2. COUNCIL DIRECTIVE 2011/70/EURATOM OF 19 JULY 2011 ESTABLISHING A COMMUNITY FRAMEWORK FOR THE RESPONSIBLE AND SAFE MANAGEMENT OF SPENT FUEL AND RADIOACTIVE WASTE

Council Directive 2011/70/EURATOM (the Euratom Waste Directive) [5], like the Joint Convention [3], requires appropriate national arrangements for a high level of safety in spent fuel and radioactive waste management. In particular, each EU Member State is required to develop a framework and a programme for the responsible and safe management of spent fuel and radioactive waste, and to implement this programme. This will ensure that an undue burden on future generations is avoided. The Euratom Waste Directive [5] is also intended to ensure adequate public information and participation in the management of spent fuel and radioactive waste. All 27 EU Member States are members of Euratom and are also Contracting Parties to the Joint Convention, in addition to Euratom itself.

## 2.3. INTERNATIONAL SUPPORTING MATERIALS

The IAEA safety standards reflect an international professional consensus on what constitutes a high level of safety for protecting people and the environment from the harmful effects of ionizing radiation. They are issued in the IAEA Safety Standards Series, which has three categories: Safety Fundamentals, Safety Requirements and Safety Guides. IAEA Safety Standards Series No. SF-1, Fundamental Safety Principles [9], presents the fundamental safety objective and principles of protection and safety, and provides the basis for the Safety Requirements, which establish the requirements to be met to ensure the protection of people and the environment. The IAEA has also published many informational publications related to safe management of spent fuel and radioactive waste, including in the IAEA Nuclear Energy Series.

Many countries are having peer reviews performed on various aspects of their spent fuel management and radioactive waste management programmes with the aim of assessing and improving their policies and practices. Such reviews are, in fact, required under the Euratom Waste Directive. For transparency, the Member States usually make these peer review reports openly available. The international organizations offer their Member States numerous expert review services related to the peaceful uses of nuclear science and technology. These peer reviews are conducted at the request of member countries and the scope of such reviews is based on the needs of the country.

There is also wide support from academia, the community, national organizations involved in the management of spent fuel and radioactive waste, etc. As a result, there are scientific and technical publications available for the wider community.

---

[2] Information on the Joint Convention [3], its latest status, documents and the results of Review Meetings are available at https://www.iaea.org/topics/nuclear-safety-conventions/joint-convention-safety-spent-fuel-management-and-safety-radioactive-waste

# 3. SOURCES OF SPENT FUEL AND RADIOACTIVE WASTE

All industrial processes result in the generation of waste that subsequently needs to be managed safely and effectively. The operation of nuclear reactors, as well as their associated fuel cycles (uranium production, enrichment, fuel fabrication and reprocessing), generates radioactive material to be managed as radioactive waste. Radioactive waste also results from the use of radioactive materials in research, medicine, education and industry. This means that all States that engage in any kind of nuclear application have to consider the production and management of radioactive waste and to make sure it is managed in a safe manner, with due regard to the level of radioactivity and in compliance with national regulations, based on IAEA safety standards.

Spent nuclear fuel is generated as a result of the operation of all types of nuclear reactors, including power reactors, research reactors, isotope production reactors and propulsion reactors. The spent fuel can be considered to be a resource for reuse or to be waste, depending on the policy and strategy of the Member State.

## 3.1. RADIOACTIVE WASTE CLASSIFICATION

The activities connected to the safe management of radioactive waste are quite different depending on the type of waste involved. As the radioactivity content of different types of radioactive waste varies greatly, the waste can be assigned to different classes. Waste is classified under national programmes according to their hazards and the available or planned management routes. Although different waste classification systems exist, the classification system used in this publication follows the definitions in para. 2.2 of GSG-1 [4], which classify waste as follows:

(1) Exempt waste (EW): Waste that meets the criteria for clearance, exemption or exclusion from regulatory control for radiation protection purposes.
(2) Very short lived waste (VSLW): Waste that can be stored for decay over a limited period of up to a few years and subsequently cleared from regulatory control according to arrangements approved by the regulatory body, for uncontrolled disposal, use or discharge.
(3) Very low level waste (VLLW): Waste that does not necessarily meet the criteria of EW, but that does not need a high level of containment and isolation and, therefore, is suitable for disposal in near surface landfill type facilities with limited regulatory control.
(4) Low level waste (LLW): Waste that is above clearance levels, but with limited amounts of long lived radionuclides. Such waste requires robust isolation and containment for periods of up to a few hundred years and is suitable for disposal in engineered near surface facilities.
(5) Intermediate level waste (ILW): Waste that, because of its content, particularly of long lived radionuclides, requires a greater degree of containment and isolation than that provided by near surface disposal. However, ILW needs no provision, or only limited provision, for heat dissipation during its storage and disposal.
(6) High level waste (HLW): Waste with levels of activity concentration high enough to generate significant quantities of heat by the radioactive decay process or waste with large amounts of long lived radionuclides that need to be considered in the design of a disposal facility for such waste. Disposal in deep, stable geological formations, usually several hundred metres or more below the surface, is the generally recognized option for disposal of HLW.

Generally, the higher the hazard, the more elaborate and/or the higher the degree of containment and isolation in the disposal concept. Depending on national polices, several waste classifications are sometimes grouped together for management in a single facility. In this case, the combined facility should be designed considering the safety of the highest class of waste it houses. The association between waste classes, activity levels and half-lives, with the boundaries between classes (shown as dashed lines) is illustrated conceptually in Fig. 1.

Some of the radioactive material has such a low content of radionuclides that the radiological impact is negligible, and it can be released from regulatory control ('clearance') in accordance with the State's regulations. This is the case for EW. Other properties, such as chemical hazards, may also affect the available management route. Some countries have a special classification ('mixed waste') that includes non-radiological hazards. An overview of national classification schemes is provided in Annex I.

Most of the radioactivity associated with radioactive waste is ILW and HLW. While VLLW and LLW comprise more than 92% of the total volume of the waste (Fig. 2), ILW and HLW typically comprise more than 95% of the total radioactivity.

## 3.2. NUCLEAR POWER PROGRAMMES

In 2019, 443 nuclear power reactors were being operated in 30 countries, generating approximately10% of global electricity. Since 2019, two new nuclear countries have emerged: Belarus and the United Arab Emirates. Bangladesh and Türkiye are currently constructing their first nuclear power plants. Italy, Kazakhstan and Lithuania have shut down their nuclear power reactors, so they do not produce any more nuclear energy (see Table 1).

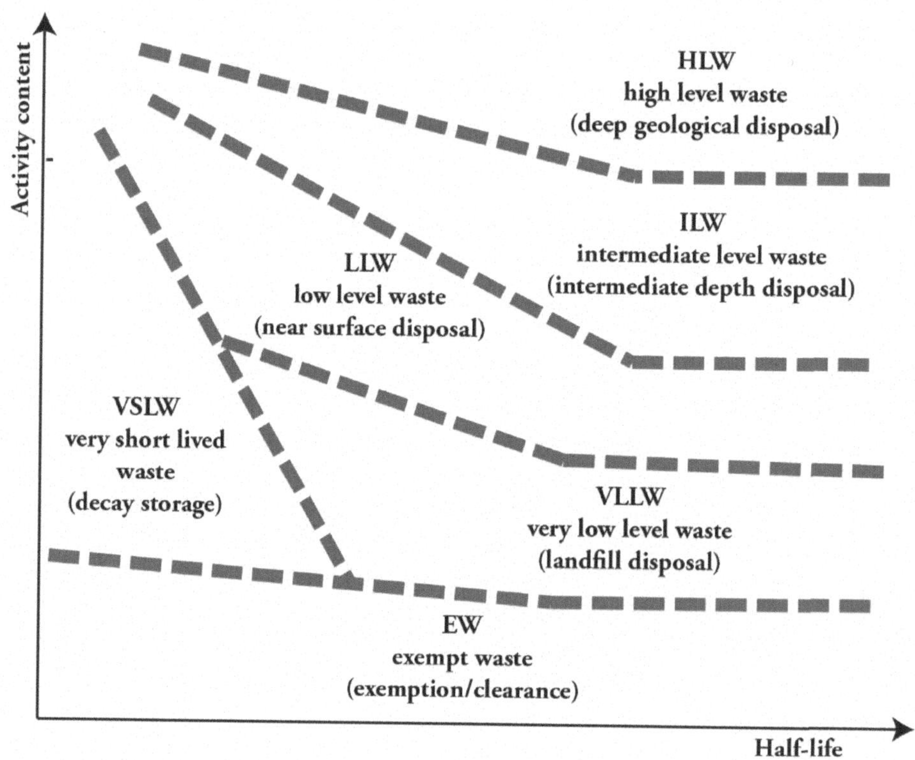


*FIG. 1. Conceptual illustration of the waste classification scheme [4].*

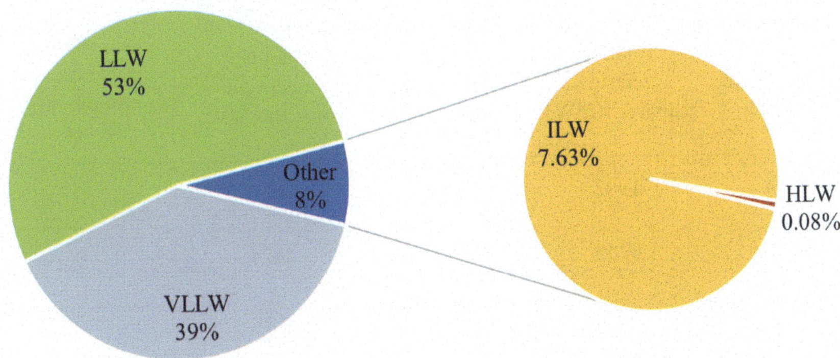


*FIG. 2. Share of different classes of radioactive waste in total volumes in storage and disposal, based on the 2019 inventory data.*

TABLE 1. IN OPERATION, UNDER CONSTRUCTION AND DECOMMISSIONING NUCLEAR POWER REACTORS, DECEMBER 2019 [10]

| Member State | In operation | | Under construction | | Decommissioning | |
| --- | --- | --- | --- | --- | --- | --- |
| | No. of units | Total net electrical capacity (MW) | No. of units | Total net electrical capacity (MW) | No. of units in the decommissioning process | No. of units decommissioned |
| Argentina | 3 | 1641 | 1 | 25 | 0 | 0 |
| Armenia | 1 | 375 | 0 | 0 | 1 | 0 |
| Bangladesh | 0 | 0 | 2 | 2160 | 0 | 0 |
| Belarus | 0 | 0 | 2 | 2220 | 0 | 0 |
| Belgium | 7 | 5930 | 0 | 0 | 1 | 0 |
| Brazil | 2 | 1884 | 1 | 1340 | 0 | 0 |
| Bulgaria | 2 | 2006 | 0 | 0 | 4 | 0 |
| Canada | 19 | 13 554 | 0 | 0 | 3 | 0 |
| China | 48 | 45 518 | 11 | 10 564 | 0 | 0 |
| Czech Republic | 6 | 3932 | 0 | 0 | 0 | 0 |
| Finland | 4 | 2794 | 1 | 1600 | 0 | 0 |
| France | 58 | 63 130 | 1 | 1630 | 10 | 0 |
| Germany | 6 | 8 113 | 0 | 0 | 22 | 3 |

| Member State | In operation | | Under construction | | Decommissioning | |
|---|---|---|---|---|---|---|
| | No. of units | Total net electrical capacity (MW) | No. of units | Total net electrical capacity (MW) | No. of units in the decom-missioning process | No. of units decom-missioned |
| Hungary | 4 | 1902 | 0 | 0 | 0 | 0 |
| India | 22 | 6255 | 7 | 4824 | 0 | 0 |
| Iran, Islamic Republic of | 1 | 915 | 1 | 974 | 0 | 0 |
| Italy | 0 | 0 | 0 | 0 | 4 | 0 |
| Japan | 33 | 31 679 | 2 | 2653 | 19 | 1 |
| Kazakhstan | 0 | 0 | 0 | 0 | 1 | 0 |
| Korea, Republic of | 24 | 23 172 | 4 | 5360 | 2 | 0 |
| Lithuania | 0 | 0 | 0 | 0 | 2 | 0 |
| Mexico | 2 | 1552 | 0 | 0 | 0 | 0 |
| Netherlands, Kingdom of the | 1 | 482 | 0 | 0 | 1 | 0 |
| Pakistan | 5 | 1318 | 2 | 2028 | 0 | 0 |
| Romania | 2 | 1300 | 0 | 0 | 0 | 0 |
| Russian Federation | 38 | 28 437 | 4 | 4525 | 6 | 0 |
| Slovakia | 4 | 1814 | 2 | 880 | 3 | 0 |
| Slovenia | 1 | 688 | 0 | 0 | 0 | 0 |
| South Africa | 2 | 1860 | 0 | 0 | 0 | 0 |
| Spain | 7 | 7121 | 0 | 0 | 3 | 0 |
| Sweden | 7 | 7740 | 0 | 0 | 6 | 0 |
| Switzerland | 4 | 2960 | 0 | 0 | 2 | 1 |
| Türkiye | 0 | 0 | 1 | 1114 | 0 | 0 |
| Ukraine | 15 | 13 107 | 2 | 2070 | 0 | 0 |

| Member State | In operation | | Under construction | | Decommissioning | |
|---|---|---|---|---|---|---|
| | No. of units | Total net electrical capacity (MW) | No. of units | Total net electrical capacity (MW) | No. of units in the decommissioning process | No. of units decommissioned |
| United Arab Emirates | 0 | 0 | 4 | 5380 | 0 | 0 |
| United Kingdom | 15 | 8923 | 2 | 3260 | 26 | 0 |
| United States of America | 96 | 98 152 | 2 | 2234 | 33 | 13 |

### 3.2.1. Spent fuel

After its use in a reactor, spent fuel is highly radioactive, emits significant radiation and heat, and is typically transferred to wet storage in a fuel pool for several years [11]. After this period (sometimes referred to as a cooling period), the spent fuel can be safely transferred to storage facilities, either wet or dry, or storage capacities in reprocessing facilities. The length of time spent fuel stays in various types of storage depends on its characteristics and intended disposition. For example, spent fuel intended to be reprocessed may spend very little time in storage (a few years), while spent fuel intended for direct disposal may spend several decades in storage, depending also on the availability of the disposal facilities.

Spent fuel contains uranium, fission products, plutonium and other heavier elements. The exact composition of the spent fuel will depend on the initial fuel type (uranium, thorium, mixed oxide (MOX), etc.) and its enrichment (i.e. percentage of fissile content), and the type and operating conditions of the reactor (e.g. thermal or fast neutron spectrum, burnup, etc.). To take advantage of the remaining fissile content of the spent fuel, some countries have adopted a closed or partially closed fuel cycle where the spent fuel is reprocessed, resulting in the extraction and reuse of the uranium and plutonium in new fuel, as well as the separation and conditioning of waste products (see Section 5.1).

### 3.2.2. Radioactive waste

During the operation of a reactor, different types of radioactive waste are generated. This waste includes filters used in water and air treatment, worn out components and industrial waste that has become contaminated with radioactive substances. This waste has to be conditioned, packaged and stored prior to its disposal. Most of this waste (by volume) has low levels of radioactivity (VLLW or LLW).

At the end of its operating life, a reactor is shut down and eventually dismantled. During dismantling, contaminated and activated components are separated, treated and, if necessary, managed as radioactive waste. The largest volumes of radioactive waste generated are in the VLLW or LLW classes. Smaller volumes of ILW are also generated. The majority of the waste (by volume) from dismantling is, however, not radioactive and can be handled as industrial waste, in accordance with the Member State's regulations.

Decommissioning of nuclear reactors and management of decommissioning waste is becoming increasingly important, as the current global fleet of power reactors is ageing. There are more than 70 power reactors that have been in operation for more than 40 years and more than 250 power reactors that have been in use for more than 30 years (see Fig. 3). It is foreseen that an increased number of nuclear reactors will be closed over the next two decades. Furthermore, as shown in Table 1, many

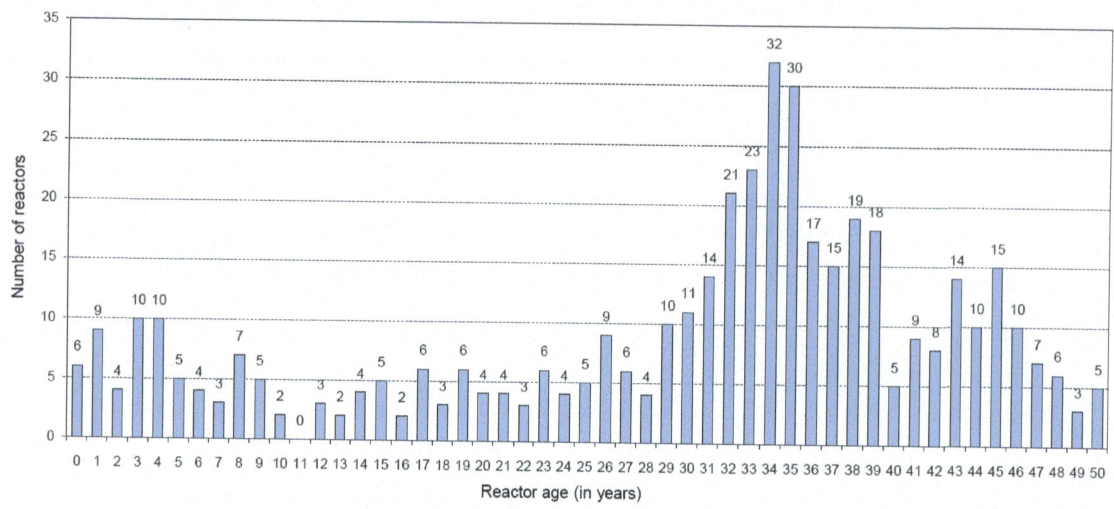


*FIG. 3. Global number of operational reactors by age (as of 31 December 2019) [10].*

nuclear power reactors have been shut down worldwide. However, currently, fewer than 20 have been completely dismantled. These have given useful experiences of complete decommissioning and handling of the radioactive components as radioactive waste. Several further units are in different stages of decommissioning, ranging from defuelling to actual dismantling. However, there are also reactors that, after removal of the spent fuel, are awaiting future dismantling while being kept in safe condition. In such cases, decontamination and dismantling might be delayed for up to 50–60 years.

Success in decommissioning may be strongly influenced by the suitability of three key elements to complete an integrated approach: a regulatory legal framework [12], the necessary provisions with regard to the funding and availability of resources, and access to technologies and experience in this field, including the presence of logistical and management solutions for the resulting materials, particularly radioactive waste.

Early planning is needed, prior to the start of actual decommissioning, to enable proper and timely treatment, conditioning, storage and disposal of the radioactive waste from decommissioning. It is important to ensure that wastes with no known safe disposition are not generated during the decommissioning process.

At different stages of the nuclear fuel cycle different materials are generated and some of them as possible waste. Since the base material, uranium, is radioactive and new radioactive elements are formed during reactor operation, radioactive waste is generated in all steps of the nuclear fuel cycle. Most of this is VLLW, LLW or ILW and is treated according the same principles as waste from nuclear reactors. The exceptions are waste from uranium (or thorium) mining and milling, which is described in more detail in Section 5.5.

Different stages of the nuclear fuel cycle result in radioactive waste and other by-products, as follows:

(a) Uranium mining and milling (UMM) generates naturally occurring radioactive material (NORM) waste. The waste rock is both the overburden rock, which contains only very low levels of NORM, and the rock from which the uranium bearing material has been separated, which contains residual uranium and other related naturally occurring radionuclides from the uranium decay chain. The mill tailing is the residue after the uranium has been extracted from the uranium bearing material to produce uranium concentrate powder or so called 'yellow cake'.

(b) Conversion of uranium oxide to uranium hexafluoride and the conversion of uranium hexafluoride to uranium oxide (as part of the enrichment process) generates VLLW and NORM waste.

(c) Enrichment generates uranium bearing waste (uranium with lower $^{235}$U enrichment levels than natural uranium). Depleted uranium (DU) is stored safely (usually in a stable chemical state), although it is

10

not always considered a waste because it can be a resource for MOX fuel, or for down-blending of high enriched uranium to lower enrichments. Several countries (e.g. France, the Russian Federation, the United Kingdom (UK)) require that the management of DU as waste be studied, if the option of reuse is not implemented on a sufficient scale to use all of the DU.

(d)  Fuel fabrication generates uranium bearing waste, which is mostly considered to be VLLW. Fuel fabrication from recycled uranium and plutonium also creates alpha bearing waste containing a range of Pu and U isotopes, as well as some of the minor actinides (Am, Np, etc.).

(e)  Reactor operation and maintenance generates a range of waste from VLLW to HLW, mostly waste with activation products created by the neutron bombardment of reactor materials (e.g. $^{60}$Co, $^{59}$Ni, $^{63}$Ni, etc.). However, owing to some fuel leakage or owing to fissions occurring outside the fuel, this waste can also contain fission products and alpha emitters. The radioactivity circulates in the primary and secondary cooling systems as well as the spent fuel storage pools, and most of it is captured by the cleanup circuits servicing these systems (e.g. filters, ion exchange systems, etc.).

(f)  Water treatment and cleaning processes in spent fuel wet storage facilities may generate filters and resins contaminated with activation products and traces of fission products and alpha emitters.

(g)  In reprocessing facilities, fission products generated in the reactor are extracted from the spent fuel and incorporated into a glass matrix (normally HLW). Claddings and structural components of the fuels are normally considered to be ILW, as well as some technological waste and effluents from the chemical processes. The latter can also be mixed with the HLW in the glass canisters. Reprocessing facilities also generate LLW and VLLW with different radiological contaminants (activation products, fission products, low levels of uranium or plutonium) as part of routine operation and maintenance activities. Manufacturing of new fuel containing recycled uranium or a uranium–plutonium MOX also generates ILW.


## 3.3.  RESEARCH, MEDICAL AND INDUSTRIAL APPLICATIONS

Research reactors are nuclear reactors used for research, development, education and training. Constructing a research reactor, as with nuclear power reactors, involves careful planning and preparation, as well as long term investment in human and financial resources. The need for funding and government oversight extends well beyond the usual timescales for a capital project of similar size. Many research reactor projects involve a wide range of interests and sometimes their missions evolve over time. Based on the Research Reactor Database,[3] there are currently more than 200 operational research reactors, approximately 80 in temporary/extended/permanent shutdown, approximately 70 under decommissioning and almost 450 decommissioned. Both the operating and decommissioning of a research reactor will produce spent fuel and radioactive waste.

While most decommissioning programmes focus on nuclear reactors of various types, all facilities that contain or handle radioactive materials will eventually have to undergo decommissioning and dismantlement. This includes research facilities, waste treatment and conditioning facilities, storage facilities, nuclear application installations, fuel cycle facilities, etc. The scale of these projects is usually smaller than for nuclear power plant decommissioning. However, some large and complex facilities, such as spent fuel reprocessing plants, can be equally, if not more, challenging to undertake. These other types of facilities tend to be much more single event type projects with little experience to draw upon. Consequently, the range of waste resulting from decommissioning of non-reactor nuclear facilities varies widely both in terms of quantity and radioactivity. Other hazards, such as chemical hazards, also require careful consideration in the planning process for these facilities.

Radiation can be used to improve quality of life in many ways, so radioactive material continues to be used broadly, both nationally and internationally, for medical, research, sterilization and other commercial applications. This means that radioactive waste is generated in a wide range of activities,

---

[3]  https://nucleus.iaea.org/rrdb/#/home

including research, the use of radioisotopes as tracers in medical and industrial applications, and irradiation of materials (e.g. for sterilization and polymerization). The typical life cycle of the radioactive material is presented in Fig. 4. Generally, the same types of treatment and handling and disposal methods are applied as for similar classes of waste resulting from nuclear power generation.

The most common fields of application for radioactive sources in industry include the calibration of measuring devices, materials testing, irradiation and sterilization of products, and level and density measurements. In medicine, radioactive sources are mostly used for radiotherapy and for irradiation of blood. The working lives of the sources used vary considerably because of the wide range of half-lives among the radionuclides used. In most countries, the devices (with the sealed source inside) are returned to the equipment manufacturer by the operator at the end of use. The source manufacturer might check for the possibility of further use of the sources and reuse parts of them. Sources that cannot be reused have to be disposed of as radioactive waste. In many countries, disused sealed radioactive sources (DSRSs) are the primary or only type of radioactive waste.

The Code of Conduct on the Safety and Security of Radioactive Sources [13] aims at helping national authorities to ensure that radioactive sources are used within an appropriate framework of radiation safety and security. The Code is a well accepted, non-legally binding international instrument and has received political support from more than 130 Member States. The Guidance on the Import and Export of Radioactive Sources [14] supplements the Code and aims to provide for an adequate transfer of responsibility when a source is transferred from one State to another. The Guidance on the Management of Disused Radioactive Sources [15] provides further guidance regarding the establishment of a national policy and strategy for the management of disused sources and on the implementation of management options such as recycling and reuse, long term storage pending disposal and return to a supplier.

NORM waste can be also produced in different industrial sectors. The most important ones include the phosphate sector, the production of titanium dioxides, water treatment, geothermal energy, the steel industry, the oil and gas industry, extraction of rare earths, etc. Depending on the classification of the waste used in the Member State, NORM might be considered as radioactive waste or not.

Many national and international government and non-governmental organizations have contributed to the increasing visibility of alternative technologies as a way to reduce risks from radioactive sources [16]. Progress with developing alternative technologies has been uneven across different applications and radionuclides, and in some applications no suitable replacement technology has been developed.


## 3.4. DEFENCE PROGRAMMES AND OTHER POTENTIAL SOURCES

Military and defence activities involving nuclear material create radioactive waste in various forms, and in some cases account for the majority of waste produced in the Member State. Neither the Joint Convention [3] nor the Euratom Waste Directive [5] requires States to report this waste. However, some States have included military and defence waste in their Joint Convention reports. The aggregated tabulations in Section 6 include any waste being declared and managed as part of the national inventory of radioactive waste.

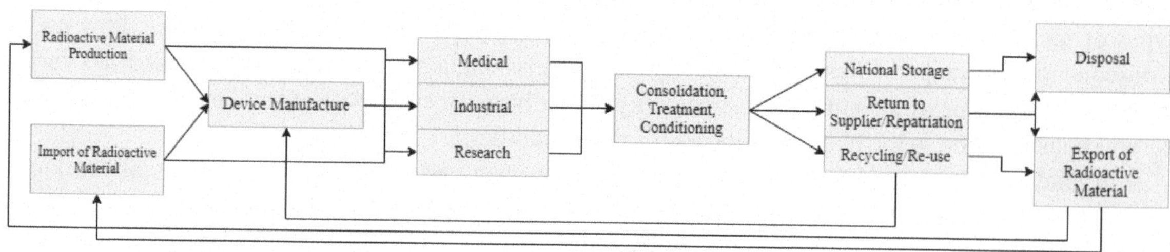


*FIG. 4. Life cycle of radioactive material.*

Other potential sources of radioactive waste include past activities that involved radioactive materials or waste generated by nuclear or industrial accidents. Usually this kind of radioactive waste presents a special challenge, as it may present an additional waste stream where the waste may range from large volume/very low activity to small volume/high activity. Waste forms may be also very variable, so waste management issues may result either from the nature of the radioactive materials (e.g. historical radium bearing waste sites) or from chemical and chemical toxic aspects. The management needs particular attention if the quantities, location and/or characteristics of the waste exceed the existing waste management infrastructures. Two major nuclear accidents in the past four decades have resulted in widespread contamination that requires significant remediation effort — Chornobyl and Fukushima Daiichi nuclear power plants. Work at both sites is currently ongoing and will likely be continuing for some time to come.

# 4. FRAMEWORKS FOR THE MANAGEMENT OF SPENT FUEL AND RADIOACTIVE WASTE

National arrangements for securing the safe management of spent fuel and radioactive waste also take into consideration international treaties and standards. A basic prerequisite, as stated in No. SF-1 [9], the Joint Convention [3] and the Euratom Waste Directive [5], is that the prime responsibility for ensuring the safety of spent fuel and radioactive waste management rests with the licence holder. It is also evident from those documents that the ultimate responsibility for ensuring that programmes are prepared for the management (including disposal) of radioactive waste rests with the State in which that waste arises. These obligations are implemented in each Member State through legislation and regulations in which the roles, responsibilities and reporting relationships of the relevant organizations are established.

## 4.1. NATIONAL POLICIES AND STRATEGIES

While the national arrangements for ensuring that spent fuel and radioactive waste are safely managed vary from country to country, there are some common features. The national legislative assembly is usually responsible for enacting legislation, which generally includes the establishment of a regulatory body and, in many cases, an implementing body for spent fuel and radioactive waste management, as well as defining the essential elements of the national policy and other related governance. Alternatively, national policy can be set out separately by governmental decree or ministerial directives. In some cases, a single policy covering both spent fuel and radioactive waste is adopted, while in other cases separate policies are issued. The IAEA's guidance on Policies and Strategies for Radioactive Waste Management [17] states that a national policy typically addresses the following:

(a) Responsibilities within the country for spent fuel and radioactive waste management;
(b) Arrangements for financing the management (including disposal and decommissioning);
(c) Preferred management options for spent fuel, policies for waste disposal, import and export of spent fuel and radioactive waste;
(d) Decommissioning of nuclear facilities;
(e) Public information and public involvement in related decisions.

To implement the national policy, one or several strategies have to be developed, which is generally the responsibility of the implementers of waste management practices, such as national waste management organizations (WMOs) (see Sections 4.4 and 4.5). In some cases, commercial entities and/or agreements with other countries are employed to implement the policy or strategy. Approval of the specific strategy by the regulatory body and/or responsible ministry is also often required.

The usual practice according to the national policy on spent fuel management and radioactive waste management is that final waste should be disposed of in the country where it is generated. This is also an expectation for Contracting Parties of the Joint Convention, as well a general requirement for EU Member States based on the Euratom Waste Directive [5]. Although spent fuel may be transferred for reprocessing to another country, the HLW or ILW from reprocessing is generally returned to the originating country for long term management.

This does not mean that countries should be precluded from fulfilling their national obligations through collaboration with other countries [18]. Some countries are seeking joint (or multilateral) solutions for the management of spent fuel and radioactive waste, including disposal in facilities that are operated jointly by, or on behalf of, several countries. The joint/multilateral disposal concepts should not be relied on as the only option for radioactive waste management in countries, owing to the uncertainties involved.

The export and import of spent fuel and radioactive waste is subject to strict controls. Many States prohibit the import of spent fuel and radioactive waste. Other States, such as France, the Russian Federation and the UK, allow the import of spent fuel from other countries, including those from research and other non-power reactors, for reprocessing services. The current practice is usually to return waste separated from recyclable materials in conditioned form to the country of origin. There are several radioactive waste processing facilities that are used by the waste producers from different countries.

States that are suppliers of sealed radioactive sources for use in medicine and industry, such as Canada, France, Germany, the Russian Federation, South Africa and the United States of America (USA), also accept the return of DSRSs.

Most countries have established national strategies for implementing radioactive waste and spent fuel management, which is in line with Requirement 1 of IAEA Safety Standards Series No. GSR Part 1, Governmental, Legal and Regulatory Framework for Safety (Rev. 1) [12] and stipulated in the Joint Convention [3]. National strategies include, for example, plans for implementing national policy, the development of the required facilities, the identification of roles and the setting of targets for the implementation of the policy. Many countries have well developed strategies and plans to manage all types of waste, from creation through to final disposal. The slow pace associated with moving towards disposal for ILW, HLW and spent fuel in many countries is dominated by the time required to perform the necessary research and site surveys, engineering and construction and to gain public acceptance of proposals to site facilities in specific areas. For these reasons, some States are implementing their chosen national strategies progressively, especially for the long term management of spent fuel and HLW. An overview of national strategies for spent fuel and different types of radioactive waste is provided in Annex II.


## 4.2. LEGAL FRAMEWORK

According to GSR Part 1 (Rev. 1) [12], the government shall establish and maintain an appropriate governmental, legal and regulatory framework for safety within which responsibilities are clearly allocated [12].

This requirement is also reflected in Article 20(2) of the Joint Convention [3]. The legal framework should include provisions to ensure sufficient and timely funding of spent fuel and radioactive waste management activities — including providing management facilities and establishing requirements for public involvement in the decision making process. While legal instruments vary, they typically assign roles and responsibilities for nuclear activities, including radioactive waste management, to operating organizations, ministries and other governmental organizations. The Country Profiles in Spent Fuel and Radioactive Waste Information System (SRIS) provide information on the national legal frameworks in each of the countries.

## 4.3. ALLOCATION OF ROLES AND RESPONSIBILITIES

Requirement 2 of GSR Part 1 (Rev. 1) [12] establishes the essential elements of a regulatory framework. At the government level, the ministries or departments of energy, industry, economy and development, with responsibilities for ensuring adequate energy supplies, often support the nuclear power industry in making arrangements for managing spent fuel and radioactive waste. The ministries responsible for ensuring that public health and the environment are adequately protected are typically responsible at the governmental level for issues related to the management of spent fuel and radioactive waste.

The role of the regulator is to ensure that nuclear activities are performed in a safe manner and in accordance with the legal and regulatory framework. For most EU and NEA members, nuclear safety regulators are now clearly separated from the national ministry in charge of energy or industry [19]. The basic responsibilities for ensuring the safety of spent fuel and radioactive waste management are assigned by all States involved in this study in accordance with the above norms, although in different ways — the differences are usually owing to variations in national legislative and regulatory systems. In some countries, for example, the owner or licence holder of a spent fuel and radioactive waste management facility is a private entity and thus is responsible for ensuring safety. In other countries, the owner or licence holder might not be completely distinct from the government and so the responsibility for ensuring the safety of spent fuel and radioactive waste management essentially rests with the State.

## 4.4. WASTE MANAGEMENT ORGANIZATIONS

Usually, the producer of radioactive waste is considered to be its first owner, so the primary responsibility for managing spent fuel and radioactive waste rests with the owner or licence holder of the facility where the spent fuel and radioactive waste originates from. At each major decision point, the implications for the safety of the management of spent fuel and radioactive waste have to be considered and taken into account. In assessing safety, several factors have to be considered, including public acceptability, cost, existing infrastructure and waste ownership.

Taking into account that the management of spent fuel and radioactive waste is considered to be a long term activity compared to the activities from which spent fuel or radioactive waste originated, there is a practical need for arrangements at the national level. Many States have created national radioactive WMOs that are responsible for developing arrangements for the disposal of spent fuel and radioactive waste. These WMOs may also be responsible for implementing management of spent fuel and/or radioactive waste (including HLW from reprocessed spent fuel). Once waste has reached WMO, the ownership of the waste might change, as it could be transferred from the producer to WMO or Member State. However, as the structure and role of WMOs in countries with a nuclear power programme varies, the ownership might stay with the producer as well.

In some countries, the generators of spent fuel and radioactive waste are responsible for all activities for safe management, encompassing the disposal of radioactive waste (including HLW from reprocessed spent fuel) and spent fuel. In such cases, the waste generators have formed WMOs that are owned and operated by them. This is true for the management of waste from nuclear power plants in Canada, Finland, Japan and Sweden, for example. In other countries, however, the State has created a separate state owned organization responsible for waste disposal (including spent fuel and/or all applicable radioactive waste classes), while the responsibility for the interim management of spent fuel and radioactive waste remains with the spent fuel or waste producer. Such an approach is used for managing all radioactive waste in China, France, Germany, the Russian Federation and Switzerland. Other countries may have a mixed approach whereby, for example, private companies are responsible for the short term management of spent fuel, whereas a State owned or State controlled body is responsible for the long term management of spent fuel and/or HLW. The private companies might also be responsible in such cases for the management of radioactive waste (with exception of HLW) and/or decommissioning.

However, some countries with small quantities of radioactive waste do not have dedicated WMOs. The nature and role of WMOs is described in Annex III.

## 4.5. FUNDING ARRANGEMENTS

The Joint Convention [3] requires that a Contracting Party has adequate financial resources available, among other things, to support the safety of facilities for spent fuel and radioactive waste management during their operating lifetime, for decommissioning and also for the activities needed for the operation/closure of disposal facilities. It is broadly recognized that those who benefit from using a source should pay for its disposition [14]; this is generally known as the polluter pays principle. This requires the establishment of a funding system for spent fuel and radioactive waste management needs, and there are different options available. The country can choose and define a scheme that is suitable for its particular needs. In most countries, spent fuel and waste producers are responsible for the funding of all activities connected to the management of spent fuel (including direct disposal if it is regarded as waste or disposal of the resulting waste if it is reprocessed) and radioactive waste (including final disposal) and for the decommissioning of the facilities.

An overview of financing schemes and funding mechanisms in different countries is given in Annex IV. The data provided in Annex IV show that funding arrangements can sometimes include the costs for management and disposal of all the radioactive waste being generated in a country, while in other cases the funding is limited to the disposal of spent fuel or HLW and the decommissioning of nuclear facilities. In the latter case, the costs of management and disposal of other types of waste are paid directly by the waste producers as an operating expense at the time when they occur. The funding arrangements described in Annex IV mainly relate to spent fuel and radioactive waste from nuclear power plants. In some countries (e.g. Finland and Sweden), this fund also covers the costs of decommissioning the facilities and managing the waste from decommissioning. In other countries (e.g. Switzerland and the USA), separate funds have been established for decommissioning.

For nuclear activities operated by the State, for example nuclear research or use of radionuclides in medicine, or for countries having historical or legacy waste, dating from past nuclear activities, the State is also most often responsible for the management of the resulting waste. In many cases, the corresponding costs are covered by the State budget (e.g. Latvia). In most cases, no segregated funding arrangement was established. Instead, the funding for current and future waste management is, and will be, met directly from government sources. In some countries, similar arrangements have been implemented for small producers of radioactive waste (i.e. the waste producers pay for waste management and disposal). In other countries, the State takes responsibility for these costs in return for fees paid by the waste producers.

Most funding systems are based on the premise that the waste producer will pay all costs for the management of the spent fuel and radioactive waste produced and that these costs will be taken from the funds that have been built up. There are different ways to collect funds. In countries with nuclear power reactors the most common method is to levy a fee per kilowatt-hour produced; however, there are also other methods for building the funds, such as establishing a target value for the fund at the end of each year. It is important to keep in mind that many of the costs associated with the management of spent fuel and radioactive waste arise long after the revenue generating activities have ceased. Therefore, it is essential to establish mechanisms to gather and protect funds during the revenue generating phase. For nuclear services provided (e.g. reprocessing or disposal), the funding can be based on a cost per cubic metre of waste delivered or per activity content. Irrespective of which mechanism is used, it is important that the actual fee is based on the best available calculated costs.

For the establishment of fees, the activities to be covered by the funding should be defined and the expected costs for the necessary facilities and activities calculated. As the timescales are large and many of these costs will occur in the future, it is important to develop a scenario and a time schedule for the management of the spent fuel and radioactive waste. As the cost calculations will inevitably often be based on very early facility designs and operation descriptions, they will involve substantial uncertainties.

There could also be uncertainties caused by the time schedule. It is also important to address the possible unexpected costs and how to secure the funds for these possible needs. The inclusion of the uncertainties can be handled in different ways, for example as a contingency or through a statistical approach. A challenge in the cost calculations is to predict how different cost types will develop in relation to general inflation, and what influence technological developments and competitiveness in the industry will have. To ensure that money will be available in the long run some countries have introduced guarantees in addition to payment of fees. The thinking is that guarantees should cover reduced income to the funds, for example owing to lower power prediction than anticipated or higher than expected cost increases. A guarantee could be in the form of a bank guarantee or a guarantee by the mother company.

The choice of margins included in the calculated costs will be dependent on the safeguards required in the funding system and the extent to which the State will take the final responsibility to cover deficiencies in funding. There might be an extra contingency applied to safeguard against unexpected costs. Alternatively, as is the case for Sweden, the fee should reflect the expected costs and unexpected costs should be covered by securities.

The collected funds can be managed in different ways. In most cases, segregated funds have been established, whereby management of the funds (i.e. collection, investment, payments) is handled by a dedicated body, often under government control. In other cases, the funds are kept inside the organization and invested, although there are also cases where the funds become part of the State budget. A key question for management of the funds is the effective return on the funds and, in this connection, what investment possibilities exist. To safeguard against cost increases owing to inflation and to keep the fees at an appropriately low level, it is important that the fund content is invested in such a way that a proper return on the money is achieved. Given that these funds are foreseen to be secured for a long time period, normally the flexibility in investments has been quite low and restricted to very secure investments, such as State or property bonds. There are cases where a certain percentage of the funds could be invested in more profitable portfolios, such as shares. The possibilities for and restrictions on the investment policy have a strong impact on the return for the funds, but they also influence the stability of the funds and the necessity of liquidity once the use of the funded money draws closer.

As the funds will exist for many decades the risk of disturbances is large. Such disturbances could include, for instance, cost increases, time schedule changes, early reactor shutdown, international and national economic turbulence, bad fund management and companies ceasing to exist. As long as the waste producing activity generates a revenue, it should be possible to adjust the funding requirements through relatively frequent recalculations of the future costs, incomes and returns. This means that changes can be accommodated through a change of the levies on the future waste generation.

In the unlikely case of insufficient funds, for example if the funds are emptied before all activities have been completed, countries' approaches differ. In some countries, the State takes over responsibility for covering unfunded costs, while in other countries the waste producer remains responsible for providing additional funding. In the latter case, the risk of insolvency of the waste producer also has to be considered, but in the extreme case it is always the State that takes the final risk. The risk of cost increases owing to disturbances will thus be taken on for the future production of nuclear electricity through increased fees and ultimately by the government. Alternatively, one could consider a system such as that in Sweden, where the obligation to pay to the fund remains even after cessation of power production. There are also some cases where the responsibility for an activity and the corresponding funds are transferred from one organization to another, such as when specialist companies take over the dismantling of reactors, as has been done in the USA. Together with the transfer of funds there is also the takeover of all the obligations connected to the activity, including paying all the costs, even if they are not covered by the funds.


## 4.6. PLANNING AND INTEGRATION

The ability of a country to deal with its spent nuclear fuel and/or radioactive waste is determined by the extent to which it has established a management policy and strategy, along with supporting infrastructure and legislative and institutional frameworks. The main objective of the system for management of spent fuel and radioactive waste is to avoid imposing an undue burden on future generations; this means that the generations that produce the spent nuclear fuel and/or radioactive waste have to develop and adequately provide for the implementation of safe, practicable and environmentally acceptable solutions for its long term management. Communication and information sharing between various organizations and various stages is important for minimization of waste generation and effective implementation.

To implement an adequate spent fuel or radioactive waste management system, a suitable degree of planning is required, starting with a basic understanding of what types of spent fuel and/or radioactive waste will arise, how much, where and when. The planning needs to be reviewed and updated periodically in order to verify that the planning assumptions are still valid and the overall management plan is still adequate and viable for the country.

Full integration of the spent fuel and/or radioactive waste management system may be challenging, especially in countries with large or complex nuclear industries or those with a long history of nuclear applications. However, there are usually opportunities to integrate the various agencies responsible for different aspects of the management system, government policy and/or regulations, long-standing practices or infrastructure, etc. A typical practice in the past was to tackle the different types of waste individually. It should be noted that, implemented in this fashion, the lowest cost solution for each individual waste stream may not be the optimal solution for the overall system. By taking advantage of possible synergies between different components of the management system and/or different waste streams, overall optimization may be achieved, leading to a reduction in costs and resource utilization. For example, a national policy could be to send all waste to a single repository, which might be a geological repository suitable for the highest level of waste in the country. If each waste class is destined for a different repository (e.g. engineered landfill for VLLW, engineered near surface for LLW and geological for ILW and HLW), then more work goes into segregating the different classes of waste. The development of alternative routes to disposal requires efficient means in terms of treatment, decontamination and characterization. This issue is important, in particular, for the management of decommissioning waste, where large volumes of VLLW or LLW will be generated. In general, a set of evaluation tools will need to be developed to support an integrated supply chain covering all aspects of waste production and management. There have been great successes in some countries, such as in the UK, in diverting a significant part of LLW from the national LLW repository to alternative disposal routes, such as licensed industrial landfill. Often the unavailability of or limitations on a disposal option and costs are good drivers to promote the minimization of radioactive waste or recycling of radioactive material, but sometimes, such as in France, the economic trade-off between direct disposal of the waste and waste treatment to reduce disposal volumes or to enable recycling can be difficult.

For countries that are just embarking on a nuclear power programme or national waste management system, there is an opportunity to build in integration right from the start [20]. The availability of existing infrastructure and resources needs to be considered in the planning. This may include collaboration or sharing of services with other countries, especially if there are suitable existing services in one of the countries that can be made available to other countries, rather than each country developing its own infrastructure. This may ultimately lead to the benefit of a much more efficient and cost effective radioactive waste management system in individual countries as well as globally.

Nuclear fuel cycle strategies need to be fully integrated with the overall spent fuel and radioactive waste management policy and infrastructure to ensure optimal use of resources. Introducing efficiencies into individual steps in isolation can create additional challenges in subsequent steps. One of the main challenges is to maintain enough flexibility to accommodate the range of potential future options for the management of spent fuel, as well as to define and address the relevant issues in storage and transportation.

Optimization of available waste management infrastructure and disposal routes should be considered because of its influence on two main decommissioning schedule drivers: time and cost. Lack of a final waste management route (i.e. disposal) should not be an excuse for postponing decommissioning or preparation for it. However, the possible implications of early decommissioning under those circumstances need to be carefully considered in order to not jeopardize future necessary actions. Experience from past decommissioning projects shows the importance of optimizing the overall waste management approach, for example via a full understanding and use of the waste management hierarchy (see Section 4.8).

## 4.7. MINIMIZATION IN THE MANAGEMENT OF RADIOACTIVE WASTE

As radioactive waste is defined in the IAEA Nuclear Safety and Security Glossary [21] as "radioactive material for which no further use is foreseen" and "that contains, or is contaminated with, radionuclides at *activity concentrations* greater than *clearance levels* as established by the *regulatory body*", it will be difficult to redefine the material once it has been declared to be radioactive waste.

Minimization of waste is "the *process* of reducing the amount and *activity* of *radioactive waste* to a level as low as reasonably achievable" (ALARA) [21]. This is "important at all stages from the design of a facility or activity to decommissioning, and is achieved by design and operations to reduce the amount of waste generated by means such as recycling and reuse, and by treatment to reduce the waste's activity, with due consideration for secondary waste as well as primary waste". Minimization principles should already have been followed in the planning stages to achieve the best results.

Minimization is a crucial part of the circular economy (Fig. 5), which returns products, parts and materials into use several times. UNIDO [22] defines the circular economy to be based on the principles that [23]:

— Products are designed to last;
— Value is maintained for as long as possible;
— Generation of waste and pollution is minimized;
— Renewable energy is used along value chains, as much as possible.

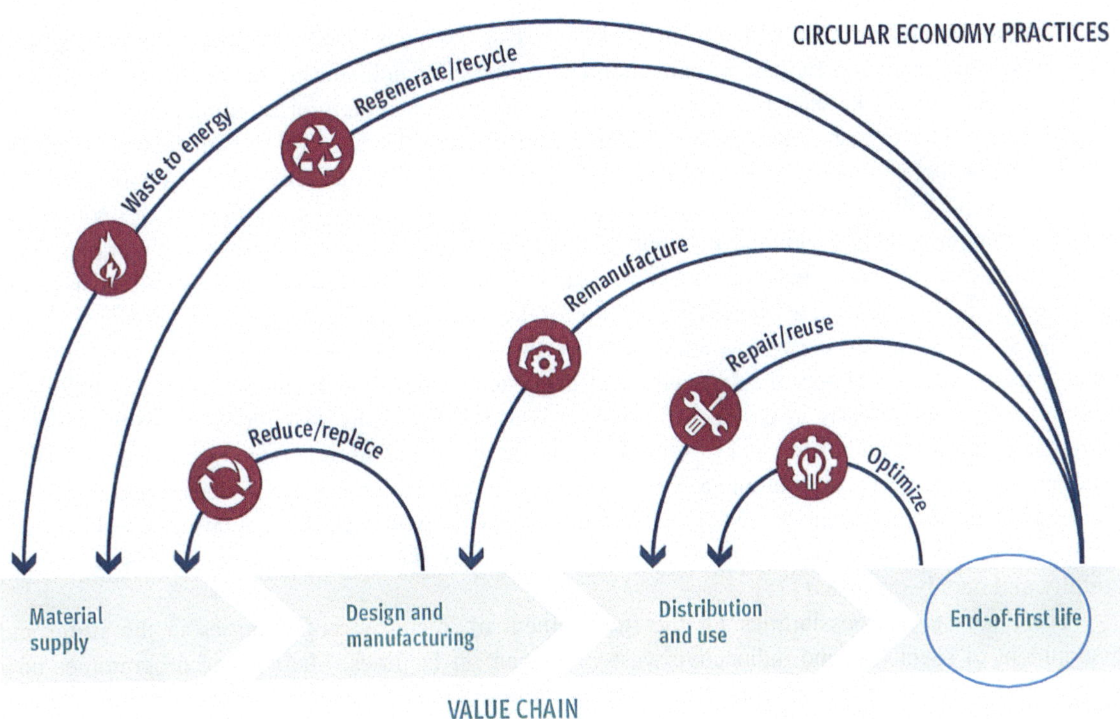


*FIG. 5. Circular economy compared to linear economy (reproduced with permission from UNIDO) [22].*

The recycling of radioactive material is a process whereby material is converted into new products, so there is a reduction of wastage of useful materials, use of raw materials and energy use. This is in line with the principles of the circular economy to extend product lifespan through improved design and servicing and relocating waste from the end of the supply chain to the beginning — in effect, using resources more efficiently by using them repeatedly, not only once.

As much as possible, everything should be reused, remanufactured or, as a last resort, recycled back into a raw material or used as a source of energy. However, there can be limited possibilities for recycling or reuse of radioactive material, especially if it is supposed to be done nationally. Reuse of radioactive material means that an item will be used again, either to perform the same function or a different function. It should be noted that there are different national approaches, depending on how the radioactive waste is defined. For example, in some countries radioactive waste is considered only as the material going to disposal. There are several other criteria to consider in minimization, such as radiation protection and public acceptance. These can weaken the advantage of waste minimization because of doses to workers during waste processing. The opportunity for recycling of materials from a nuclear facility can also be restricted or reduced because of opposition from the public.

This means that for minimization in different phases of spent fuel and radioactive waste management there are different factors that need to be considered:

— Regulatory and licensing issues: compliance of the option with the applicable regulation;
— Technical and operational issues: availability of technology and facilities to process waste;
— Safety and ALARA issues;
— Economic and scheduling issues: costs, duration for implementation of solutions, compatibility with agenda of waste generation;
— Public acceptance and stakeholder issues.


## 4.8. INVOLVEMENT OF INTERESTED PARTIES

Stakeholder involvement, which is an integral part of a stepwise process of decision making, may take the form of sharing information, consultation, dialogue, or deliberating on decisions at different phases. It should always be seen as a meaningful part of formulating and implementing good policy.

It is important to secure stakeholder involvement through the life cycle of all nuclear facilities, including spent fuel storage facilities and final radioactive waste disposals. International experience has shown that, especially in the case of disposal facilities, the project's progress often relies upon public support. Decision making on long term spent fuel and radioactive waste management is complex, as it not only concerns the current generation, but also future ones since the disposal facilities are designed to operate for many decades and to contain the hazard for thousands of years.

The stakeholders' expectations should be taken into consideration through different activities and interactions in order to enhance the satisfaction of interested parties [24]. It can be useful and helpful to involve the community early in the decision making process. This helps to build mutual trust between operators, government authorities and stakeholders, especially among the general public. Increased public participation in decisions can promote a greater degree of understanding of the issues related to nuclear power and spent fuel/radioactive waste management, especially with regard to actual risks and benefits. Public confidence is improved when issues that are raised by the public are taken seriously and are carefully and openly evaluated [25].

There are several possibilities for the involvement of interested communities in the siting and development of spent fuel and radioactive waste management facilities. Many siting programmes now incorporate local partnerships and there are even examples of waste management implementing bodies proposing to involve local stakeholders in joint studies, and in the interpretation and review of ongoing site investigation, assessment of the potential impacts on human health and the environment, and development of plans for monitoring these issues during facility operation and closure [26].

The typical steps for implementing stakeholder involvement programmes [24] can be listed as follows:

— Develop a strategy for stakeholder involvement;
— Develop plans for implementing this strategy;
— Ensure that the capacity to effectively implement these plans is available;
— Implement these plans;
— Continually monitor the effectiveness of these actions and look for ways to improve.

To earn trust from stakeholders, it is also important to have licensing authorities that are competent, as well as being independent from political and industrial influence in their decision making and deliberation. Many regulators already incorporate public comment sessions in their licensing and review processes. It is also crucial to establish clear criteria for the decision making process, so that it is understandable how and when, for example, the facility siting process can move from one step to the next. There needs to be clarity on the scope for decision making and identification of the point in the process where specific decisions are finalized and are not subject to being revisited [27].

Although decision making processes vary considerably by Member State, depending on culture, history and governmental structure, stakeholder involvement is worthy of consideration. Stakeholder involvement is an essential component of various international conventions and treaties, most commonly related to the strategic environmental assessment and environmental impact assessment. The Convention on Access to Information, Public Participation in Decision-Making and Access to Justice in Environmental Matters, otherwise known as the Aarhus Convention [28], has more than 40 Contracting Parties and is not only an environmental agreement, but also covers government accountability, transparency and responsiveness. It grants the public rights and access to information. The Convention on Environmental Impact Assessment in a Transboundary Context, otherwise known as the Espoo Convention [29], has 45 Contracting Parties and sets out the obligations of the Parties to assess the environmental impact of certain activities at an early stage of planning and make sure that the stakeholders are involved in the process. It also lays down the general obligation of States to notify and consult each other concerning all major projects under consideration that are likely to have a significant adverse environmental impact across boundaries. At the same time the Euratom Waste Directive [5] ensures the provision of necessary public information and participation in relation to spent fuel and radioactive waste management while having due regard to security and proprietary information issues.

# 5. SUMMARY OF CURRENT STRATEGIES, PRACTICES AND TECHNOLOGIES

The previous sections covered the sources of spent fuel and radioactive waste, as well as the frameworks for their safe management. This section gives an overview of the current strategies, practices and technologies for safe management of spent nuclear fuel and radioactive waste. Other properties, such as chemical hazards, may also affect the available management route.

The preferred strategy for the management of all radioactive waste is to contain it (i.e. to confine the radionuclides to within the waste matrix, the packaging and the disposal facility) and to isolate it from the accessible biosphere [29]. Disposal is defined as intentional emplacement of the waste in a facility without the intent to retrieve that waste. Disposal options are designed to contain the waste by means of passive engineered and natural features and to isolate it from the accessible biosphere to the extent necessitated by the associated hazard. Figure 6 illustrates the disposal options based on the classes of radioactive waste [30].

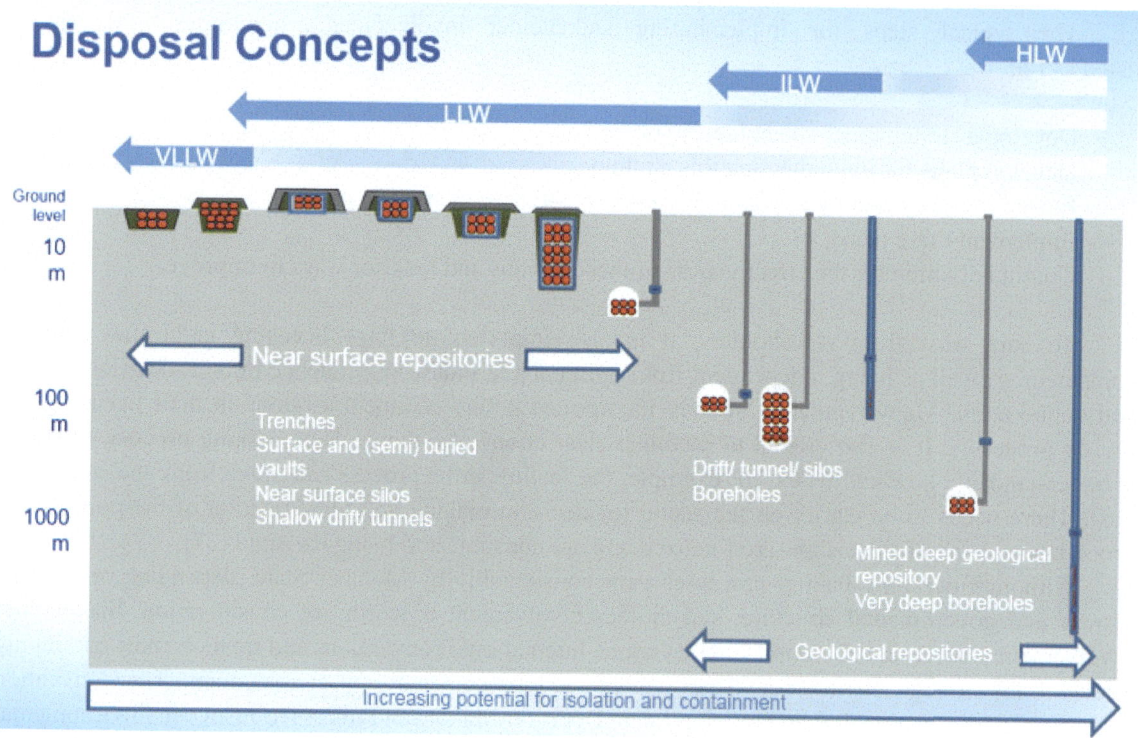


*FIG. 6. Conceptual illustration of disposal concepts for different classes of radioactive waste [30].*

## 5.1. SPENT FUEL AND HIGH LEVEL WASTE

Spent fuel is generated from the operation of nuclear reactors of all types, including research, isotope production, power production, district heating and propulsion reactors. By volume, HLW forms <1% of the global volume of radioactive waste, but it consists of approximately 95% of the total activity of the radioactive waste. The activity level of HLW is high enough that heat generation should be considered in the design of the waste management facilities. In countries where spent fuel is considered to be waste, it is classified as HLW.

### 5.1.1. The 'open cycle' and the 'closed cycle'

The currently envisaged strategies to ensure safe and cost effective overall management of spent fuel differ from one country to another and can be described as follows:

(a)   The 'open cycle' or 'once through" or 'direct disposal' strategy, in which spent fuel is considered as waste;
(b)   The 'closed cycle' (including the 'partially closed cycle') strategy, in which the spent fuel is considered to provide a potential future energy resource.

In the open cycle option, spent fuel is stored for several decades to allow the decay heat to be reduced and also to optimize the operation of the deep geological disposal facility. After a period of storage, the spent fuel will be encapsulated in a robust corrosion resistant container to meet disposal acceptance criteria and will be disposed of in a deep geological repository.

In the closed cycle option, the spent fuel is reprocessed to recover valuable fissile materials (uranium, plutonium). In reprocessing, spent fuel is separated into its main components: uranium, plutonium and HLW (containing minor actinides, fission and activation products). The HLW contained in spent fuel is then stored to allow the decay heat to be reduced, pending future disposal, normally in a deep geological repository (DGR). The uranium and plutonium can be recycled as nuclear fuel for reactors, while the minor actinides, fission and activation products are currently considered to be waste products. The minor actinide and fission product waste are conditioned in a stable vitrified matrix and stored in a very stable matrix purposely designed for storage, transport and disposal. The main activation products (hulls and end pieces of fuel assemblies) are compacted and conditioned in steel canisters, stored pending future disposal compatible for ILW. The spent fuel might go through one or more cycles of reprocessing in order to recover valuable material.

Currently, the countries that operate large scale reprocessing facilities are France, India and the Russian Federation (Fig. 7). The UK formerly operated two reprocessing facilities. In 2022 all reprocessing facilities in UK ceased operation. China is operating a pilot plant and is looking to deploy an industrial facility. The reprocessing facility in Japan (Rokkasho-mura plant) is currently under safety review and planning to resume operation. Other countries, including Belgium, Bulgaria, the Czech Republic, Germany, Hungary, Italy, Japan, the Kingdom of the Netherlands, Slovakia, Spain, Sweden, Switzerland and Ukraine, have used services provided by foreign facilities (in the UK, France and the Russian Federation (including during the time of the former USSR)) for the reprocessing of their spent fuel.

The commercial capacity for reprocessing was 3500 t of heavy metal per annum in December 2019 (Table 2). The end of reprocessing operations in the UK in 2022 is reducing the worldwide availability of reprocessing capacity until new facilities in China, Japan and the Russian Federation come into operation.

Spent fuel reprocessing in another country is subject to strict controls and is performed based on commercial contracts under the umbrella of bilateral national agreements. In most cases, these commercial contracts provide that the waste produces, as well as the valuable fissile material (usually in the form of fuel for recycling), together with the conditioned HLW from spent fuel reprocessing (as well as fuel component compacted waste in some cases), are sent back to the country from where the spent fuel originated.

FIG. 7.  Maintenance cell at ORANO, France (courtesy of Eric Larrayadieu).

TABLE 2. COMMERCIAL SCALE REPROCESSING FACILITIES (AS OF 31 DECEMBER 2019)

| Member State | Facility | Capacity (t HM/a) | Status |
| --- | --- | --- | --- |
| France | UP2-400, La Hague | 400 | Under decommissioning |
| | UP2-800, La Hague | 800 | In operation |
| | UP3, La Hague | 800 | In operation |
| | UP1, Marcoule | 600 | Under decommissioning |
| Japan | Rokkasho-mura | 800 | In commissioning |
| Russian Federation | RT-1, Mayak | 400 | In operation |
| | RT-2, Zheleznogorsk | 60 | Under construction |
| UK | NDA THORP, Sellafield | 900 | Ceased operation in 2018 |
| | NDA Magnox Reprocessing, Sellafield | 1500 | Ceased operation in 2022 |

Reprocessing spent fuel using aqueous separations results in high level liquid waste that is typically vitrified; that is, conditioned to produce a chemically durable and heat and radiation resistant engineered solid matrix waste form. Several types of glass (e.g. borosilicate, phosphate) and some ceramics are used for the treatment and conditioning of HLW. The glass containing waste is poured into containers which are also used for storage. The vitrification process and all container handling operations are performed remotely in shielded cells. Significant experience has been obtained with the vitrification process in Belgium, France, Japan, the Russian Federation, the UK and the USA. The HLW is then stored for some decades to allow levels of heat generation to be reduced, in a similar way as for spent fuel. Following storage, HLW is to be disposed of in a DGR (see Section 5.1.4). Some countries include fuel cladding and structural material that was separated during reprocessing within the HLW class.

The uranium separated during reprocessing, the so called RepU, can be recycled as fuel in present day reactors following conversion and re-enrichment if necessary. Recycling RepU from the reprocessing of LWR fuel in a pressurized heavy water reactor, such as a Canada deuterium–uranium (CANDU) reactor, has also been developed in China. As an example, natural uranium equivalent fuel is an innovative fuel designed to work in synergy with current and planned spent fuel reprocessing technologies in China. It blends RepU from LWRs with deuterium–uranium to create natural uranium equivalent fuel powder that is used to fabricate pressurized heavy water reactor fuel [31].

The separated plutonium can be recycled into MOX fuel, in which DU and plutonium oxides are combined. MOX fuel has been used for decades in LWRs worldwide and in a few Generation IV reactors (France in the past and today in the Russian Federation), where the energy value of the uranium and plutonium can be better utilized. At present, only one country, France, has a commercial MOX fuel fabrication facility in operation for manufacturing of LWR fuel. This facility, MELOX, has provided MOX fabrication services since 1995 for France and several other countries [32]. Belgium and the UK formerly operated MOX facilities, called Belgonucléaire and SMP, respectively, which ceased operation in 2006 and 2011. MOX plants are also planned to come into operation over the next few years, as is the case in Japan. Additional multirecycling options in LWRs are under development, such as the regenerating mixture (REMIX) fuel currently in the demonstration phase in the Russian Federation (recycling all the uranium and plutonium without separating them and topping up with some fresh uranium enriched to a higher level than 5%), as well as some other fuel concepts under study in France (MOX2). Several countries implementing or considering reprocessing today have plans, at different stages of development, for future fast neutrons reactors, although at present only the Russian Federation operates such reactors at a commercial scale (BN-600 and BN-800).

In addition, there are ongoing initiatives to use DU as fuel or to recycle all recovered long lived actinides together (i.e. with plutonium) in fast reactors. This strategy would make it possible to increase utilizations of the uranium in nuclear fuel from less than 1% to well over 90%, which would result in waste containing mainly short lived fission products, reducing the waste disposal burden. New reprocessing technologies are being developed to be deployed in conjunction with fast neutron reactors that will burn all long lived actinides, including all uranium and plutonium, without separating them from one another.

A summary of the current fuel cycle strategies adopted in different countries is given in Table 3. It shows that the majority of countries have adopted the open cycle or use it for referencing, while the countries with some of the largest nuclear programmes (e.g. China, France, India, Japan, the Russian Federation) have adopted the closed cycle. Some countries with a small nuclear fleet, such as the Kingdom of the Netherlands, have also opted for the closed cycle strategy, with reprocessing services provided by one or more of the larger countries with this capability. Table 3 also shows that although several countries have chosen the open or closed cycle, there are also countries that are keeping their options open.

## TABLE 3. NUCLEAR POWER FUEL CYCLE STRATEGIES

| Member State | Commercial scale reprocessing facility | | Spent fuel currently in another country for reprocessing | Earlier reprocessing, but practice currently ceased | Planning direct placement of spent fuel in a repository | Keeping options open |
|---|---|---|---|---|---|---|
| | Existing | Planned | | | | |
| Armenia | | | | | | ✓ |
| Argentina | | | | | | ✓ |
| Belarus | | | ✓ | | | |
| Belgium[a] | | | | ✓ | ✓ | ✓ |
| Brazil | | | | | | ✓ |
| Bulgaria[a] | | | ✓ | | | |
| Canada | | | | | ✓ | |
| China[b] | ✓ | | | | ✓ | |
| Czech Republic[a] | | | | ✓ | ✓ | |
| Finland | | | | ✓ | ✓ | |
| France | ✓ | | | | | |
| Germany | | | | ✓ | ✓ | |
| Hungary[a,c] | | | | ✓ | ✓ | |
| India | ✓ | ✓ | | | | |
| Italy | | | ✓ | | | |
| Japan[d] | ✓ | | ✓ | | | |
| Korea, Republic of | | | | | | ✓ |

TABLE 3. NUCLEAR POWER FUEL CYCLE STRATEGIES (cont.)

| Member State | Commercial scale reprocessing facility | | Spent fuel currently in another country for reprocessing | Earlier reprocessing, but practice currently ceased | Planning direct placement of spent fuel in a repository | Keeping options open |
|---|---|---|---|---|---|---|
| | Existing | Planned | | | | |
| Lithuania | | | | | ✓ | |
| Mexico | | | | | ✓ | |
| Netherlands, Kingdom of the | | | ✓ | | | |
| Pakistan | | | | | | |
| Romania | | | | | ✓ | |
| Russian Federation | ✓ | ✓ | | | | |
| Slovakia | | | | ✓ | ✓ | ✓ |
| Slovenia | | | | | ✓ | ✓ |
| South Africa | | | | | | ✓ |
| Spain | | | | ✓ | ✓ | |
| Sweden[e] | | | | ✓ | ✓ | |
| Switzerland | | | | ✓ | ✓ | |
| Türkiye | | | | | ✓ | |
| United Arab Emirates | | | | | ✓ | |
| UK | | | | ✓ | ✓ | |
| Ukraine[f] | | | ✓ | ✓ | | ✓ |
| USA | | | | ✓ | ✓ | |

[a] Mixed policy: some fuel has been or will be reprocessed and other fuel will be or may be disposed of directly.
[b] The main policy in China is domestic reprocessing. However, it is planned to dispose of some fuel, mainly from CANDU reactors, directly.
[c] Earlier fuel returns to the Russian Federation, but no requirement to return waste from reprocessing to Hungary.
[d] A commercial scale facility has been constructed at Rokkasho-mura and is undergoing test operation.
[e] Earlier reprocessing was performed abroad.
[f] Some spent fuel is sent to the Russian Federation for reprocessing. Other fuel is stored awaiting a final decision.

### 5.1.2. Transport of spent fuel and high level waste

The management of spent fuel and HLW involves a number of transport steps between nuclear power plants or other facilities using nuclear fuel, storage facilities, encapsulation/packaging facilities and/or reprocessing facilities, as well as eventually disposal facilities [33]. Most transport operations are performed within one country, but some journeys cross national frontiers. For countries reprocessing their spent fuel but having no reprocessing facilities of their own, such transboundary movements are necessary. Similarly, the transboundary movement of spent fuel is necessary for countries sending spent fuel from research reactors and other reactors back to the country of origin of the fuel.

Transport is typically undertaken in specially designed transport containers that provide security, shield workers and the general public, and perform other nuclear safety functions, such as managing decay heat, ensuring subcriticality and providing neutron shielding [34]. These transport operations are strictly controlled according to national regulations, which are meant to be based on the transport regulations in IAEA Safety Standards Series No. SSR-6 (Rev. 1), Regulations for the Safe Transport of Radioactive Material [35]. Each State involved in a transboundary movement has to take the appropriate steps to ensure that the transport operation is undertaken in an appropriate manner and with the authorization of the countries of origin, destination and transit.

### 5.1.3. Storage

After spent fuel has been discharged from the reactor, it is usually stored for some time in a water filled spent fuel pool to cool it and provide shielding from its radiation. The length of the storage period varies from a few years up to several decades, depending on the spent fuel management strategy adopted. Usually when spent fuel is recycled, the storage period is generally relatively short — a decade or less. In countries that have decided on a direct disposal option or that have yet to make a decision, the storage period can be much longer. Storage systems include wet storage in storage pools or dry storage in storage casks, canisters or vaults built for the purpose. An overview of the types of storage used in different Member States is provided in Table 4.

All nuclear power reactors have spent fuel storage pools for the initial decay heat cooling storage period upon discharge from the reactors. They were included in the original design of the reactors. Additional storage capacity, wet or dry, can be built to provide additional storage capacity as needed. The new storage facilities are built outside the containment building and are known as away from reactor (AFR) stores, and they can be either inside or outside the boundaries of the nuclear power plant.

Access to an AFR site may require transport over public roads, railways, sea lanes, etc. AFR facilities are typically purpose built, under a separate licence, for spent fuel storage located away from the main reactor buildings or site. They can be dedicated to one or multiple reactors or they can be a centralized facility serving more than one nuclear power plant. The storage technology for new AFR stores was initially wet storage (Figs 8 and 9), but dry storage techniques of different types have been developed (Fig. 10) and are now widely adopted. Reprocessing facilities are normally equipped with large AFR pools at the reception for buffer storage before reprocessing. More information about types of spent fuel storage, as well as examples of them in use, can be found in the IAEA's Guidebook on Spent Fuel Storage Options and Systems (Technical Reports Series No. 240) [11].

The canisters for HLW (produced during reprocessing of spent fuel) are stored in air cooled vaults or similar casks to those used for spent fuel storage. Each reprocessing plant has large vaults for canister storage — mainly for its national HLW. In Germany and Switzerland, HLW is stored in casks, while Belgium, Japan and the Kingdom of the Netherlands use dry vault storage technology (e.g. the HABOG facility in the Kingdom of the Netherlands) [36].

The IAEA's Guidebook on Spent Fuel Storage Options and Systems [11] provides guidance to Member States on spent fuel storage options, describing the history and observed trends of spent fuel storage technologies, gathering operational experiences and lessons learned, describing the evolving aspects related to higher burnup and mixed oxide (MOX) spent fuel, and the extension of the storage time frames.

## TABLE 4. SPENT FUEL STORAGE, AS AT END OF 2019

| Member State | Spent fuel storage type |
| --- | --- |
| Argentina | Wet and dry storage |
| Armenia | Wet and dry storage |
| Belgium | Wet and dry storage |
| Brazil | Wet storage |
| Bulgaria | Wet and dry storage |
| Canada | Wet and dry storage |
| China | Wet and dry storage |
| Czech Republic | Wet and dry storage |
| Finland | Wet storage |
| France | Wet storage, wet storage at reprocessing plants before reprocessing |
| Germany | Wet and dry storage |
| Hungary | Wet and dry storage |
| Japan | Wet and dry storage, wet storage at reprocessing plants before reprocessing |
| Kazakhstan | Dry storage |
| Korea, Republic of | Wet and dry storage |
| Lithuania | Wet and dry storage |
| Mexico | Wet and dry storage |
| Netherlands, Kingdom of the | Wet storage before transport to France for reprocessing |
| Romania | Wet and dry storage |
| Russian Federation | Wet and dry storage<br>Wet storage at reprocessing facilities before reprocessing |
| Slovakia | Wet storage |
| Slovenia | Wet storage |
| South Africa | Wet and dry storage |
| Spain | Wet and dry storage |
| Sweden | Wet storage |

TABLE 4. SPENT FUEL STORAGE, AS AT END OF 2019 (cont.)

| Member State | Spent fuel storage type |
| --- | --- |
| Switzerland | Wet and dry storage |
| Ukraine | Wet and dry storage |
| UK | Wet and dry storage |
| USA | Wet and dry storage |

FIG. 8.  *Wet AFR storage of spent fuel at Clab, Sweden (courtesy of SKB).*

*FIG. 9. The dry storage hall at ZWILAG Zwischenlager Würenlingen AG, Switzerland (courtesy of ZWILAG).*

### 5.1.4. Disposal

There is a broad consensus among technical experts that the preferred method of ensuring long term safety for spent fuel and HLW is isolation in a DGR. Geological disposal facilities for long lived waste will provide passive multibarrier isolation and containment of radioactive materials. Emplacement in carefully engineered structures buried deep within suitable geological formations provides the long term stability typical of a stable geological environment [36, 37]. Countries that need to dispose of their spent fuel and HLW are studying different available geological media for it [37, 38].

In the case of the open cycle option, before being sent to the DGR the spent fuel will have to be encapsulated in a corrosion resistant and mechanically stable container, which will provide isolation for a suitable duration. The vitrified HLW waste form in a stainless steel canister is specifically designed for long term durability in storage and disposal. In some countries an additional corrosion resistant overpack is also considered. The requirements for the container life and integrity depend on the DGR concept and the chosen geological medium.

Posiva Oy in Finland applied for an operational licence for an encapsulation plant and geological disposal facility in Onkalo site. The construction of underground access tunnels at the ONKALO facility in Finland started in December 2016 and the construction of the encapsulation plant started in 2019. The first construction phase of the deposition tunnels has been completed (five deposition tunnels excavated). In Sweden, the Government approved the 'allowability' of the construction of an encapsulation plant and disposal facility for spent fuel in January 2022. In France, the detailed concept for an HLW and ILW disposal project for Cigéo (Industrial Centre for Geological Disposal) was achieved in 2020 and the environmental impact assessment was issued in 2021. In 2022 it was recognized as the final solution for management of HLW and ILW. It is expected that the construction licence for HLW deep geological disposal will be submitted in 2022.

There are formal site selection processes underway in several other countries, such as Canada, Germany and the UK [37]. In countries with both spent nuclear fuel and HLW for disposal, a single DGR for both materials is typically the adopted approach. Most other countries with spent nuclear fuel are working towards national solutions, although they are largely at the early planning stage. Some countries have also indicated an interest in developing multinational disposal facilities, in addition to their own national programme.

Research related to the DGR option has been undertaken for several decades using a range of underground research laboratories (URLs). These URLs have an important role in waste disposal programmes and are also valuable in building confidence in national programmes [39–41]. Currently, there are approximately 20 URLs in use, for example HADES in Belgium, KURT in the Republic of Korea and Krasnoyarsk URL in the Russian Federation.

### 5.1.5. Spent fuel from non-power reactors

The amount of spent fuel from non-power reactors is much smaller than that from nuclear power reactors. Fuel from non-power reactors, however, raises some specific challenges, as it sometimes has higher enrichment than, and a different composition from, power reactor fuel. Although non-power reactors are often built in countries with a nuclear power programme, these reactors are also operated in countries without nuclear power plants, where non-power reactor fuel is one of the most important factors in waste management. The IAEA works with Member States to develop a variety of nuclear education and training programmes, one of which is the Internet Reactor Laboratory. This is a cost effective virtual reactor that provides the possibility to use research reactors remotely, so Member States without an existing research reactor can develop their nuclear infrastructure.

At present, most non-power reactor spent fuel is returned to the country of origin of the fuel, mainly the Russian Federation and the USA, and thus does not require disposal in the country where it has been used. A few countries, such as Australia, Belgium and Sweden, have decided to reprocess either part or all of their spent non-power reactor fuel. Some countries have to consider disposal of this spent fuel nationally, and this might be a challenging task, especially if they do not have a nuclear power programme. The IAEA Nuclear Energy Series NW-T 1.11, Available Reprocessing and Recycling Services for Research Reactor Spent Nuclear Fuel [42] provides an overview of the available reprocessing and recycling services for non-power reactor spent fuel.

## 5.2. INTERMEDIATE LEVEL WASTE

ILW generally contains significant amounts of long lived radionuclides and therefore requires disposal at depths that provide containment and isolation from the biosphere over the long term. ILW requires shielding during handling and storage. It should be noted that the definition of ILW used in GSG-1 [4] is used throughout this publication, which means that the ILW covered in the publication includes all forms of ILW that require a greater degree of containment and isolation than near surface disposal can provide.

### 5.2.1. Processing

The processing of ILW either takes place at the facilities where it is generated or at a purpose built facility (which can also be a centralized facility). Processing consists of the collection, segregation, decontamination, volume or size reduction and stabilization prior to packaging [21]. Drying, evaporation, high pressure compaction, melting and cementing are common technologies applied in the treatment and conditioning of ILW [39]. Care needs to be taken during treatment to make sure that radioactivity concentrations will not increase beyond the capability of the treatment facilities or packaging to handle the resulting radiation levels and the extent of heat emission.

Depending on its intended storage or disposal destination, ILW is often treated and conditioned by incorporating it into a matrix (e.g. cement) within a suitable container to provide the required radiation shielding [40]. In some cases, where additional matrices are not required to ensure safety, conditioning is limited to packaging. In other cases, the waste object itself (such as a large vessel with internal contamination) forms the container, once suitably sealed.

Concrete containers with steel reinforcement, steel drums and steel boxes are commonly used for waste packaging. Their dimensions are selected to meet safety requirements and to be compatible with the dimensions of transport casks and disposal vaults. ILW containers can either be self-shielded or rely on external shielding to provide the necessary radiation protection. Both design concepts are used extensively.

### 5.2.2. Storage

After processing, storage of the product is often necessary if suitable disposal facilities are not available. Storage for periods of up to 100 years or longer can be considered as an option provided that the waste containers will remain intact and are not subject to degradation. Attention needs to be given to the provision of adequate containment and shielding. Heat removal may also be required in some cases, although not to the same extent as for HLW.

### 5.2.3. Disposal

The only licensed disposal facility for long lived ILW is the Waste Isolation Pilot Plant (WIPP), USA, where long lived, non-heat generating waste from defence activities is disposed of in a geological repository built in salt beds. Elsewhere, ILW is held in storage until a disposal facility suitable for this material becomes available. Germany and Switzerland envisage that all LLW and ILW will be disposed of in one multipurpose deep geological facility, obviating the need to separate waste containing short and long lived radionuclides before disposal. The Schacht Konrad facility is licensed and under construction in Germany. In France, long lived ILW will be disposed together with HLW in the planned Cigéo (Centre industriel de stockage géologique/Industrial Centre for Geological Disposal) facility.


## 5.3. LOW LEVEL WASTE

Taken together, VLLW and LLW typically account for more than 95% of the volume but less than 2% of the radioactivity of all radioactive waste. LLW does not generally require significant shielding during handling and interim storage. The waste is suitable for disposal in engineered near surface facilities. However, some countries, such as Germany and Switzerland, are implementing policies that do not foresee separate disposal facilities for every radioactive waste class, and therefore their LLW might be disposed of at deeper facilities.

### 5.3.1. Processing

As with ILW, the treatment and conditioning of LLW either takes place at the facility where it is generated or at a purpose built facility (which can be a centralized facility). The waste is segregated, treated, conditioned, packaged, monitored and stored, as appropriate, before being transferred to the disposal facility. Drying, incineration, evaporation, high pressure compaction, melting and cementing are common processes applied to the conditioning of LLW [42]. Concrete containers, steel drums and steel boxes are commonly used for waste packaging. Subject to meeting all relevant safety requirements, their dimensions are selected to fit the dimensions and shapes of disposal spaces and transport packages.

### 5.3.2. Storage

The options for storage of LLW are broadly similar to those for ILW (see Section 5.2.2). Storage for longer periods can be considered as an option, provided that the waste containers remain intact and are not subject to degradation. However, as LLW disposal facilities are available in a number of Member States (see Section 5.3.3), the storage periods for LLW can be quite short.

### 5.3.3. Disposal

LLW, most of which has a half-life of less than 30 years, is disposed of in near surface repositories in many countries (see Annex II and Figs 10–12). These are trenches or concrete vaults into which containerized waste is placed. An engineered cover system is placed over the waste to limit water infiltration and surface erosion and to prevent intrusion by humans or burrowing animals. The facilities are subject to surveillance until the hazard associated with the waste has declined to acceptable levels (typically a few hundred years). While disposal of LLW in a near surface facility is a typical strategy for many countries, some (e.g. Canada, Finland, Germany, Hungary, the Republic of Korea, the Kingdom of the Netherlands, Sweden and Switzerland) have chosen, or are considering, the option of disposing of LLW in repositories at depths between 50 m and 1000 m. These facilities should not require long term surveillance. Several countries have both licensed and operated geological disposal facilities for LLW, either as purpose built facilities (e.g. Bátaapáti repository in Hungary; Gyeongju facility in the Republic of Korea; Himdalen repository in Norway; SFR, Final Repository for Short-Lived Radioactive Waste, in Sweden) or as converted facilities from former mines of various types (e.g. Richard in the Czech Republic, Asse II and Morsleben in Germany, Baita Bihor in Romania).

A small number of countries are considering co-location of LLW in geological facilities with ILW, HLW or spent fuel. Co-disposal can result in a simpler waste management system because fewer facilities need to be developed. However, co-location can also introduce design complexity to avoid interferences between the waste types (e.g. decomposition of LLW can result in the generation of complexing agents that reduce the safety of higher level waste), as well as significant increases in the volume of material requiring handling at geological depths.

## 5.4. VERY LOW LEVEL WASTE

VLLW often exists in large volumes and is mainly generated during the decommissioning of nuclear facilities or from the cleanup of contaminated sites. Typical VLLW includes concrete, soil and rubble. This class is currently only recognized as a distinct classification by a small number of States (e.g. France, Japan, Lithuania, Spain and Sweden). In most other countries' classification systems, it is included as part of the LLW stream.

### 5.4.1. Processing

VLLW is typically not subject to extensive processing, apart from its packaging, owing to the very large quantities involved and the low content of radionuclides. In countries where the clearance concept is used, the volume of potential VLLW can be reduced by appropriate characterization to separate those components that can be released from regulatory control as cleared waste.

### 5.4.2. Storage

Generally, VLLW is stored at the site of its generation or in a centralized storage facility until it can be transported to a suitable disposal facility. During this stage, a simple shelter or temporary cover might be sufficient to provide protection from wind, rain, etc.

*FIG. 10. LLW disposal facility in Olkiluoto, Finland (courtesy of TVO).*

### 5.4.3. Disposal

In France, Slovakia and Spain, VLLW is disposed of in purpose built disposal facilities in shallow trenches with engineered covers, often near the site of generation to avoid the transport of large volumes of material (Fig. 13). Sweden and Lithuania developed an above ground design using a concrete slab. In other countries it is disposed of together with other waste types, such as LLW, or in countries such as the UK with hazardous waste.

## 5.5. URANIUM MINING AND MILLING WASTE, NORM WASTE

Uranium mine waste rock and mill tailings are normally managed and disposed of close to the uranium mine or the uranium mill. The waste rock and tailings are not packaged but rather are contained in nearby locations with suitable barriers (stable mounds with an appropriate cover system) to minimize their radiological and non-radiological impact on the surrounding environment. In some countries, uranium mining and milling (UMM) tailings and in situ leaching waste are not classified as radioactive waste. Hence, these countries do not report the waste as radioactive waste, while others do. UMM can

*FIG. 11.  LLW disposal facility in El Cabril, Spain (courtesy of Enresa).*

*FIG. 12.  LLW disposal vault in Bátaapáti, Hungary (courtesy of PURAM).*

*FIG. 13. VLLW disposal at the Cires facility, France (courtesy of ANDRA).*

then be classified as long lived VLLW or in some cases LLW. Uranium extraction by the in situ leaching method usually also generates smaller volumes of radioactive waste and different waste forms.

Radioactive residues are also generated from the oil and gas industries (e.g. scales and sludges), mining of other minerals and products (e.g. residues from extraction of thorium and rare earth elements), and the treatment and usage of drinking and process water.

If NORM is classified as radioactive waste, depending on the national waste management concept, this is usually considered to be VLLW or LLW. NORM waste is not specifically discussed in this publication, although some countries have reported NORM waste in SRIS.

## 5.6. DISUSED SEALED RADIOACTIVE SOURCES MANAGEMENT

Sealed radioactive sources are widely used in medicine, industry and agriculture and, because of this, they are found in almost all countries. For many countries, these are the only radioactive materials to be handled, and they require storage and eventually disposal. The life cycle of a sealed radioactive source is presented in Fig. 14.

The management of disused sealed radioactive sources is covered in the IAEA Safety Standards Series No. GSR Part 3, Radiation Protection and Safety of Radiation Sources: International Basic Safety Standards [43] and by the Joint Convention [3]. However, because of the special nature of disused sealed radioactive sources (DSRSs) and their widespread use, specific international standards have been developed for their management, including the following:

— Code of Conduct on the Safety and Security of Radioactive Sources [13];
— Council Directive 2013/59/Euratom of 5 December 2013 laying down basic safety standards for protection against the dangers arising from exposure to ionizing radiation, in particular Section 2 on control of radioactive sources [44];
— Council Regulation 1493/93/Euratom of 8 June 1993 on shipments of radioactive substances between Member States [45].

The Guidance on the Import and Export of Radioactive Sources was published in 2012 [14], and in 2018, supplementary Guidance on the Management of Disused Radioactive Sources [15] was developed and published, providing more detail on the effective management of DSRSs.

Depending on the intended use, sealed radioactive sources include a wide variety of radionuclides and activity levels. The Code of Conduct [13] and GSR Part 3 [43] categorize radioactive sources according to their potential to cause serious health effects (Table 5).

At some point, sealed sources have to be replaced, usually because their activity has declined to a level below which they are no longer suitable for their intended purpose. They are then considered to be 'spent' or 'disused' sources. DSRSs are either managed together with other waste in a category commensurate with their hazard (e.g. LLW, ILW, HLW) or separately, again in a manner commensurate with their hazard. An overview of the national strategies used in DSRS management can be found in Annex II.

There has been significant progress regarding the management of disused sealed sources. While some Member States have a well established regulatory framework for disused sealed source management, others are still facing challenges in that regard. Especially with regard to historical waste, orphan sources or sources imported before the enforcement of the return policy remain a challenge for some States. Many Member States consider disused sealed sources to be waste, while other recycle and reuse them. However, the safety of long term management of disused sealed sources is still considered to be a challenging issue.

One major issue is to create an inventory and a follow-up system in countries to provide a good knowledge of what sealed sources are used and who the users are. The management of orphan and disused sealed sources remains overarching, as it has been considered during several Review Meetings of the Joint Convention. In addition, a collection system for disused sealed sources with a (financial) motivation for the user to use this system has to be implemented. There are several international initiatives for securing the safe storage of disused sealed sources. Recognizing the need to assist Member States in the safe and effective and management of disused sources, the IAEA has focused on the development of a series of publications dealing with the handling, conditioning, storage and disposal of such sources. Guidance on the management of DSRSs, including problems encountered and lessons learned, is included in the IAEA Safety Standards Series No. SSG-45, Predisposal Management of Radioactive Waste from the Use of Radioactive Material in Medicine, Industry, Agriculture, Research and Education [46] to support well informed progress and decisions in managing disused sources.

TABLE 5. CATEGORIES OF SEALED RADIOACTIVE SOURCES [23]

| Category | Risk in being close to an individual source | Examples of uses |
| --- | --- | --- |
| 1 | Extremely dangerous to the person | Radioisotope thermoelectric generators<br>Irradiators |
| 2 | Very dangerous to the person | Industrial gamma radiography sources<br>High/medium dose rate brachytherapy sources |
| 3 | Dangerous to the person | Fixed industrial gauges<br>Well logging gauges |
| 4 | Unlikely to be dangerous to the person | Bone densitometers<br>Level gauges |
| 5 | Most unlikely to be dangerous to the person | Permanent implant sources<br>Lightning conductors |

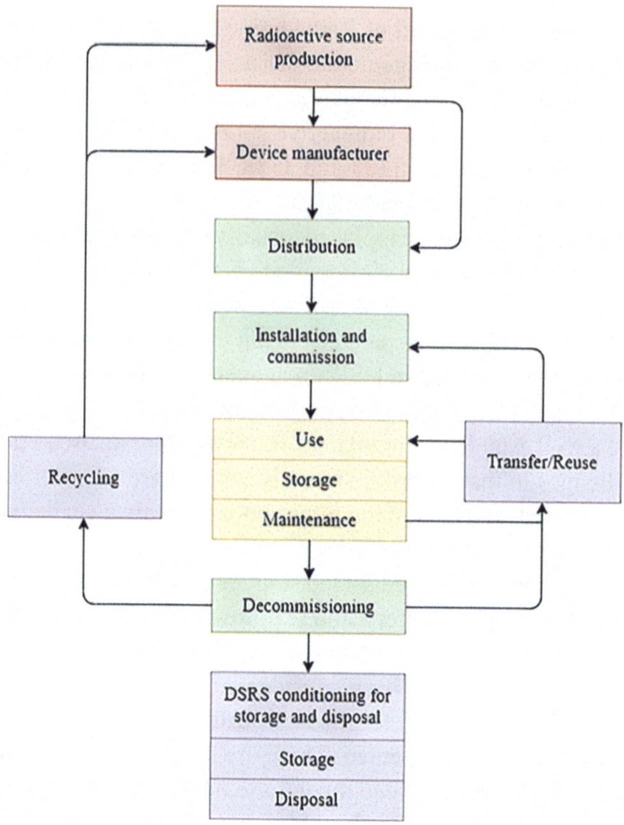


*FIG. 14. The life cycle of a sealed radioactive source.*

### 5.6.1. Storage and conditioning

States with nuclear power facilities are likely to have the capacity for long term storage or disposal of DSRSs, together with other types of radioactive waste. For many small countries, however, safe and secure storage or disposal of the sources presents an ongoing challenge. The management practices for DSRSs are very similar to those for LLW and ILW.

Sources with short half-lives (e.g. $^{192}$Ir, half-life of 74 days) can be stored until the radioactivity in the source decays to low enough levels to allow release from regulatory control (i.e. clearance), while others (e.g. $^{226}$Ra, which until recently was widely used) remain potentially hazardous for tens of thousands of years. Where disposal options are not available, long term storage facilities are required for many types of DSRS. Effective management involves repackaging the source, checking the condition of the source or source container regularly, and providing appropriate safety and security measures.

### 5.6.2. Return to supplier, reuse and recycle

Recycling for further use is the preferred option for managing disused sealed sources. If the preferred option is not possible, the option is to return it to its supplier [46]. As a result of the challenges associated with disposing of DSRSs safely, especially in countries with little or no radioactive waste management infrastructure, current good practice is to return the sources to the manufacturer for refurbishment, recycling or storage/disposal. A number of countries insist upon this as a condition of the import and sale of sealed sources within their territory.

There are many programmes to collect disused sources for reuse or recycling, or for transfer to another licensee. Typically, a prior arrangement with the user, and in some cases a 'one for one' exchange, where the user returns a disused source and concurrently purchases a replacement source, is needed to return a source to the manufacturer or supplier. These kinds of practices are very common in some application areas, for example with industrial radiography, panoramic irradiators and other irradiators that use $^{60}$Co, teletherapy and brachytherapy.

Several arrangements can be agreed between the user and supplier. Often the used source is replaced with a new source, and there can be no cost associated with the supplier or manufacturer taking the used source back. However, the return of a disused source to a manufacturer or supplier can be challenging even when a returns programme is in place. One of the challenges can be related to the documentation requirement, for example when the user needs to provide documentation concerning the country of origin of the source and where manufacture occurred. Other challenges can arise in relation to the fact that the source manufacturers may have production facilities in several countries and different components of a source may be manufactured in facilities in different countries, making it difficult to determine which country the disused sealed radioactive source should be returned to. There are other challenges — limited availability of certified transportation containers, required certification for the sources, etc. [16].

Recycling is an effective way to delay the actual disposal of a source until another option becomes available. Even when both reuse and recycle have been implemented effectively, this still only involves a small percentage of the large number of sources that require disposal.

Various recycling methods are available, such as recovery of the sealed source or transmutation by linear accelerator. Production of $^{225}$Ac for cancer therapy by photon induced transmutation of $^{226}$Ra is an example of recycling. Recycling reduces the amount of radioactive material that needs to be produced; however, at the same time it should be taken into account that these actions have to be cost effective and technically feasible.

While the return of the DSRSs to the supplier is a widely used option, it is not always possible, as the original supplier may, for example, be unknown or no longer exist, or the means of transport, regulatory framework or financial resources may not enable transportation of the sources.

Recycling of DSRSs is always a technically demanding task that requires particular expertise and authorization.

Most States have the strategy of returning disused sealed sources to the manufacturer and the country of origin or are in the process of developing centralized facilities. An overview of the national management strategies for disused sealed sources is provided in Annex II.

### 5.6.3. Disposal

If no further use is foreseen and it cannot be otherwise removed from regulatory control, the only sustainable long term option is disposal. As such, disused sources for which no recycling or repatriation options exist should be declared to be radioactive waste and should be managed as such, in compliance with relevant international legal instruments, safety standards and good practices.

For those disused sources that cannot be returned to a supplier or reused and that cannot be stored until they decay to clearance levels, disposal is the final step in their management. Some countries, particularly those with a nuclear power programme, may have the option to co-dispose of their disused sources in a near surface or geological disposal facility. It will, however, need to be verified that the sources comply with the waste acceptance criteria set up for those facilities.

Where co-disposal is not possible, disposal in one or more boreholes may offer a solution. Disposal in boreholes offers a safe and secure disposal solution. The concept of borehole disposal of disused sources has been studied extensively and developed over the last two decades. In that time, it has evolved from a conceptual idea into a mature disposal solution. Today, projects for the borehole disposal of DSRSs are ongoing in Ghana and Malaysia and are being considered in several other countries.

# 6. INVENTORIES

## 6.1. DATA SOURCES

The main sources of information used for inventories and forecasts are Spent Fuel and Radioactive Waste Information System (SRIS) and publicly available Joint Convention National Reports. In the case of some EU members, their openly available National Reports on the implementation of the Euratom Waste Directive have been used as well. The National Profiles, Joint Convention National Reports and EU National Reports cover almost 92% of all nuclear power plants in the world. This provides a good basis for making regional and global aggregations of waste volumes. Nuclear power plants generate significant quantities of spent fuel and radioactive waste, and by comparison, countries without nuclear power plants generally have much smaller amounts of radioactive waste (and of spent fuel if they operate research reactors).

The estimation of the global inventory depends on the availability of the inventory data for countries with nuclear power plants. All countries have some radioactive waste, which in many cases entirely comprises DSRSs. The aggregated numbers have been rounded for presentation purposes and the United Nations country groupings have been used for regional aggregation of the data.

### 6.1.1. Spent Fuel and Radioactive Waste Information System

Spent Fuel and Radioactive Waste Information System provides a good structure for data collection, which could also be used for aggregation of data at regional or global levels and facilitates data analysis. The data fields include, inter alia, amounts of spent fuel and radioactive waste, together with transformation matrices, to enable the transfer of the volumes given in national classification systems to the classification in GSG-1 [4]. A total of 19 Member States have submitted data to SRIS[4].

### 6.1.2. Joint Convention National Reports

Joint Convention National Reports are produced triennially and Contracting Parties to the Joint Convention are encouraged to publish their National Reports.[5] Data have been used from this source for Member States that did not submit data to SRIS.

The Joint Convention National Reports include listings of spent fuel and radioactive waste management facilities in the Contracting Parties. They also provide inventories of spent fuel and radioactive waste in the States based on their national waste classification systems. Forecasts of spent fuel and waste storage and disposal are, however, not provided. National Reports vary in the level of detail provided and in the measurement units used, requiring translation of the waste quantities presented according to national classification into the equivalent GSG-1 waste classification.

### 6.1.3. Euratom Waste Directive National Reports

The Euratom Waste Directive requires submission of a triennial report to the European Commission on the implementation of the Directive. These reports should give a comprehensive but concise high level overview of how a Member State complies with the Directive, with an emphasis on major changes and progress made since the previous report [47]. The inventory dates and timing of these reports are similar to the Joint Convention National Reports for several Member States.

---

[4] See sris.iaea.org

[5] See www-ns.iaea.org/conventions/results-meetings.asp

EU Member States are required to provide an inventory of all spent fuel and radioactive waste and estimates for future quantities in their national programme, including those from decommissioning, clearly indicating the location and amount.

## 6.2. DESCRIPTION OF DATA AGGREGATION

As noted previously, the precise definition of what constitutes radioactive waste and its classification into levels or categories varies widely among countries, which creates some inherent difficulties in aggregating inventory data regionally and globally. This section describes the approach taken for data collection and the model used to aggregate data from different countries into a common framework.

### 6.2.1. Conversion to IAEA waste classification

This publication uses the waste classification in GSG-1 [4] to present global values. This classification is based on the disposal route required to provide long term safety. However, some States intend to dispose of waste in facilities normally reserved for waste that presents a greater long term hazard (e.g. disposing of LLW in a geological repository). Furthermore, the boundaries between different classes are not defined by quantitative activity levels but instead depend on the safety case for a specific facility. On that basis, waste of a particular class in one country might not have precisely the same level of activity as the same class of waste in another country — although the differences at the margins are typically not significant. Note that the total amount of all classes of waste remains the same and can be stated with a good degree of accuracy, and only its allocation between different classes or categories may be subject to variation.

To assist the conversion to the waste classification in GSG-1 [4], while reporting inventories in SRIS, the respondents were asked to include a conversion matrix, indicating the proportions of waste from a national class corresponding to the appropriate waste classification in GSG-1. In case of data taken from the publicly available National Reports under the Joint Convention [3], estimates were made for a conversion matrix if the national waste classification differs from the waste classification in GSG-1 [4].

### 6.2.2. Constraints in determining global inventory

The data presented in SRIS and the Joint Convention National Reports has provided the basis for preparation of the global and regional aggregated data presented in this publication. However, this process involves some additional uncertainties. For example, the waste volumes can also be presented in different ways and the determination of 'as disposed' waste volumes requires assumptions to be made, which inevitably involves the use of approximations concerning the waste processing and disposal strategies. This step tends to be particularly complex in the case of liquid waste.

The recognized gaps and uncertainties in the estimation of global inventory data include the following:

(a) Lack of data on some countries. This will result in an underestimate of total inventories.
(b) Uncertainties in the translation of data from national waste classification systems to the waste classification in GSG-1 [4] for aggregation purposes. This will affect the distribution of waste volumes among the various waste classes (VLLW, LLW, ILW, HLW) but will not affect the overall total amount of waste.
(c) Differences in the way that various States report waste volumes (e.g. 'current as stored' volumes versus forecast 'as disposed' volumes, use of actual physical volume of waste packages, versus the volume envelope it might occupy in a repository). This will affect the reported volumes of waste. On average, however, the overall effect on accuracy of the global inventories should not be significant owing to offsetting increases and decreases as well as rounding of the aggregate numbers.

(d)   Different reporting dates. These will affect the accuracy of a 'snapshot' for a given date. However, most of the reporting dates are within a year or two of the selected reference date for this publication (31 December 2019). Given that in most cases the accumulated waste and spent fuel volumes do not grow very quickly and given the very large residual inventories in countries with large programmes, the overall effect on the accuracy of the global inventories should not be significant.

(e)   Different approaches to the clearance of radioactive waste.

(f)   The inclusion of unprocessed liquid waste in the totals. In some cases, no distinction was made in country reports between unprocessed liquid waste versus solid waste. The potentially large volumes of liquid waste can distort the overall data if not accounted for separately. Therefore, where a country has distinguished between liquid and solid waste, either by direct statement or inference from a waste classification, liquid waste quantities are handled separately from solid waste quantities in all relevant tables of this report.

The project supporting the development of this publication did not include a quantitative analysis of the level of uncertainty in the presented information. Lessons learned in the collection and analysis of data for this publication will be incorporated into later phases, and modifications will be sought to improve accuracy and to minimize uncertainties in the aggregated data.

### 6.2.3.   Conversion to disposal volumes

The input for SRIS requests that waste volumes indicate both the current state of the waste and its anticipated volume for disposal. To help minimize inconsistency in volumes, this publication uses 'as disposed' volumes where available, followed by 'as stored' when only this has been reported. Estimations are, however, necessary to calculate the as disposed volume, taking into account the repository requirements and the conditioning and packaging plan.

For some countries, with a known or assumed conditioning and disposal route, it is possible to transform the storage volume to the disposal volume. For States without established plans for a repository and corresponding waste package geometries, several assumptions need to be made, for example concerning what further conditioning and packaging will be required for 'disposal ready' packages. In such situations, greater uncertainty may exist concerning the disposal volumes. This uncertainty might increase if there is a possibility that conditioned waste packages are eventually placed in larger containers or overpacks for disposal.

The Status and Trends project includes an initiative by the IAEA, the NEA and the EC to harmonize the spent fuel and radioactive waste inventory data reported to the different agencies for various purposes to reduce the reporting burden on Member States and to ensure the consistency of data reported. It was agreed that the volumes of conditioned waste ready for disposal should be used. This is also recommended by the European Nuclear Safety Regulators Group [47]. However, it should be noted that there is no universal agreement on the definition of 'disposal volume'. For any given country, the definition and/or calculation method is usually embedded in regulations, national policy or a facility licence.

### 6.3.   CURRENT INVENTORIES OF SPENT FUEL

The spent fuel inventories provided in this publication do not distinguish between fuel that is considered to be a waste in the responding State and fuel that is considered to be an asset (i.e. intended to be reprocessed). The global totals include all countries where information is available. The data include spent fuel from nuclear power plants, demonstration and research reactors and other kinds of reactors (e.g. isotope production). The amount of spent fuel is presented in tonnes of heavy metal (t HM) and describes the mass of heavy metals (e.g. plutonium, thorium, uranium, minor actinides) contained in the spent fuel.

It should be noted that spent fuel that has been sent for reprocessing but has not yet been reprocessed is included in the amount of spent fuel currently in storage in the country to which it has been sent. Additional data on historical amounts of spent fuel that have been reprocessed have been extracted from other sources, such as the annual reports from commercial reprocessing facilities. Spent fuel that has been reprocessed is no longer in the form of fuel but has been separated into various types of waste and recyclable components. The unit of measure for waste from the reprocessing of spent fuel is the cubic metre and is included as part of the LLW, ILW and HLW as appropriate. Fuel that has been reprocessed is included separately in the tables under 'sent for reprocessing' to give a total for all the spent fuel that has been produced since the beginning of the nuclear power age.

### 6.3.1. Nuclear power plant spent fuel

The aggregation in Table 6, as well as those in subsequent tables and charts, gives a global summary of inventories. Subtotals for United Nations geographical regions and for the Member States of the Joint Convention, EU and NEA are provided in addition to the global total. These subgroupings allow the reader to see the aggregated amounts for various Member State groupings. Details for individual countries can be found in the Country Profiles located in the annexes.

About one third of all spent fuel discharged from nuclear power plants has been sent to be reprocessed, remaining is stored, pending processing or disposal. Most spent fuel is held at nuclear power plant sites in wet storage in the reactor pools. Fuel inside the reactor core is not included in the inventory, since it is not considered to be spent until it has been discharged from the core.

After initial storage for cooling for at least a few years in the reactor pool, some spent fuel has been transferred to dry storage or to centralized wet storage facilities. The total amount of spent fuel in storage was about 302 000 t HM as of the end of 2019. Figure 15 shows the share of spent fuel stored either in dry or wet storage [11].

TABLE 6. REPORTED SPENT FUEL IN STORAGE, AS OF 31 DECEMBER 2019

| Region | Wet storage (t HM) | Dry storage (t HM) | Not specified (t HM) | Total (t HM) |
|---|---|---|---|---|
| Africa | 1100 | 50 | n.a.[a] | 1150 |
| Americas | 78 800 | 71 100 | n.a. | 150 000 |
| Asia | 36 400 | 8100 | 8500 | 44 500 |
| Europe | 47 200 | 15 400 | 102 700 | 201 500 |
| Oceania | n.a. | n.a. | n.a. | 1 |
| Global total | 163 500 | 94 600 | 33 800 | 302 000 |
| Joint Convention Contracting Parties | 163 500 | 94 600 | 33 800 | 302 000 |
| EU Member States | 37 200 | 13 000 | 4 000 | 60 600 |
| NEA Members | 144 000 | 85 200 | 33 800 | 269 300 |

[a] n.a.: not applicable.

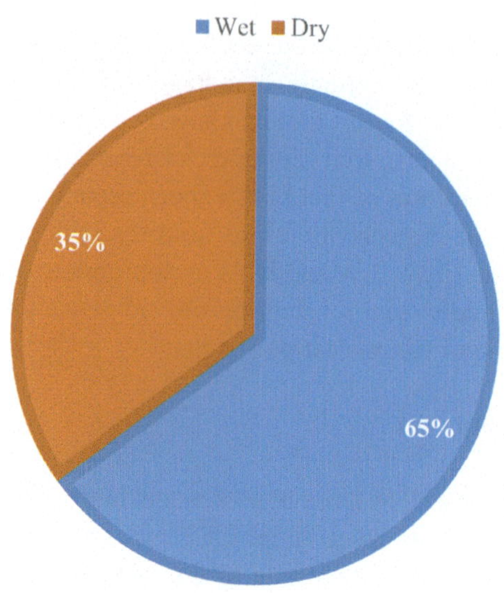

*FIG. 15. Nuclear power plant spent fuel storage by type.*

### 6.3.2. Spent fuel from research and other reactors

A number of States operate non-power reactors, such as research, isotope production, experimental, prototype or propulsion reactors. The national inventories of spent fuel from these reactors are summarized in Table 7. It is noteworthy that these amounts are less than 1% of the amount of spent fuel that originates from nuclear power plants. It should also be noted that the fuel quantities from non-nuclear power plant reactors are generally not publicly reported to the same level of detail as for nuclear power plants. However, a typical research reactor has a core capacity in the order of a few kilograms of uranium fuel, whereas a commercial nuclear power plant might have a core of 100 tonnes or more. Isotope production reactors may have a capacity of a few tonnes.

There can also be a difference in the enrichment of the fuel in $^{235}$U. Many research reactors and isotope production reactors were originally designed to operate using high enriched uranium (HEU) fuels, whereas power reactors utilize low enriched uranium fuels (roughly 0.7–5%), leading to a potential difference in the amount of uranium that is discharged in the spent fuel. This also depends on the amount of 'burnup' of the $^{235}$U. In many cases, the research reactors have been converted to operate with low enriched uranium. For example, the policy of the USA has been the minimization, and ultimately elimination, of HEU in civilian research reactors worldwide since 1978 [48].

The majority of the spent fuel in storage from non-power reactors is in North America. This is because spent fuel from prototype power reactors is placed in the research reactor category in Canada and the USA. Most spent fuel from research and other reactors in many countries has been returned to suppliers for reprocessing or disposal (usually the Russian Federation or the USA), and in these cases that spent fuel will become part of the inventory of the receiving country.

## 6.4. CURRENT INVENTORIES OF RADIOACTIVE WASTE

Most of the radioactivity present in radioactive waste (up to 95% of the total) is present in HLW (including spent fuel, when declared as waste). In terms of volume, the situation is reversed and more than 95% of the total volume of waste comprises LLW or VLLW. The hazard presented by any toxic agent is a complex combination of the quantity, the particular chemical components and their respective concentrations in the waste (in this case mainly the radionuclides), the physical and chemical form of the waste, the radioactivity level and the exposure scenarios. Generally, chemical and physical forms of waste that are mobile in the environment are more hazardous. Limiting its mobility is therefore an important

TABLE 7. REPORTED SPENT FUEL STORED FROM RESEARCH AND OTHER REACTORS, AS OF 31 DECEMBER 2019

| Region | Wet storage (t HM) | Dry storage (t HM) | Not specified (t HM) | Total (t HM) |
|---|---|---|---|---|
| Africa | 0 | 0.1 | n.a.[a] | 0.1 |
| Americas | 34 | 2437 | n.a. | 2471 |
| Asia | 159 | n.a. | 300 | 459 |
| Europe | 184 | 54 | 73 | 311 |
| Oceania | n.a. | n.a. | n.a. | 0 |
| Global total | 377 | 2491 | 373 | 3241 |
| Joint Convention Contracting Parties | 377 | 2491 | 373 | 3241 |
| EU Member States | 16 | 23 | 73 | 112 |
| NEA Members | 376 | 2490 | 73 | 2939 |

[a] n.a.: not applicable.

reason for conditioning waste prior to disposal, as well as for selecting suitable geology when siting a disposal facility.

Solid waste and liquid waste are described separately. This differentiation is important because of the significant volume of liquid waste and the large volume reduction achievable from processes such as evaporation, filtration and vitrification, among others, depending on the chemical composition and amount of water. Typically, liquid waste is processed for solidification soon after it has been generated, rather than placing it in storage. In States that follow this approach, only a small part of the national radioactive waste inventory exists in liquid form. In some countries, a past practice was to store some waste in liquid form with the intention of processing and converting it to solid form at a later stage and, as a result, the waste still exists in this form at many sites.

Most radioactive waste is either in storage awaiting the development of a suitable disposal facility, awaiting further treatment pending disposal in a licensed facility, or has already been disposed of. In general, only solid waste is placed into disposal facilities, although past practices in some countries included direct injection of liquid waste into underground formations for disposal. This strategy is still practised in the Russian Federation and is also practised in non-nuclear industries in many countries, such as for the disposal of waste from oil and gas extraction, which may contain important concentrations of naturally occurring radionuclides.

Disposal is defined as intentional emplacement in a facility without the intent to retrieve. Although some States require the possibility of retrieving the disposed waste for some period of time after disposal, this is still considered to be disposal for the purposes of this publication.

### 6.4.1. Solid radioactive waste

Solid waste includes inherently solid materials, such as metals, plastics and other dry materials, as well as solidified liquids. In the case of unprocessed waste, and in some States, 'solid waste' can also include small amounts of liquids or 'wet solids' (such as filter cake or dewatered ion exchange resins). Solid radioactive waste includes disused sealed radioactive sources if the Member State considers them as radioactive waste. Figure 16, based on the National Profiles, shows the global totals for different types of solid radioactive waste in storage and disposal, as of 31 December 2019. The data shown in Fig. 16 represent as disposed volumes, based on the conversion matrices provided by respondents (see Section 6.2.3 for a discussion of the uncertainties inherent in such an approach).

It is evident that the majority of the volume of waste consists of VLLW and LLW. VLLW has a smaller volume than LLW because often this category of waste was created more recently in a country's classification system and was not retroactively applied to radioactive waste that had already been disposed of. As noted above, most of the radioactivity is contained in the much smaller volumes of ILW and HLW. For VLLW and LLW, the majority of the waste generated has already been disposed of. For ILW and HLW, however, the majority of the waste so far generated is currently in storage awaiting the development of appropriate disposal facilities.

Table 8 summarizes the volume of solid radioactive waste in storage and Table 9 summarizes the volume of solid radioactive waste in disposal, as of 31 December 2019. The values for the waste classes are as defined by the individual countries and the totals have been rounded to the nearest 1000 m³.

It can be seen that 85% of the already generated VLLW and LLW has been disposed of, as disposal options are available for most of the generated VLLW and LLW. For ILW, the fraction already disposed of is ~10%, and for HLW there is no disposal option available. Since the 2013 edition of this publication, the fraction of disposed VLLW and ILW has increased, and the fraction of disposed LLW has decreased.

### 6.4.2.  Liquid radioactive waste

While most States process liquid radioactive waste into solid form within a short time of it being generated, a few — notably, the Russian Federation and the USA — have large volumes of liquid waste in long term storage. Much of this waste results from defence activities and is only now being dealt with through the design, construction and licensing of liquid waste treatment facilities.

Generally, amounts of liquid waste are given as stored volumes. This generally does not include liquids held in short term storage awaiting processing. It is difficult to estimate the final as disposed volumes, since this will largely depend on the eventually selected processing and conditioning methods. In most cases, this will result in a very large reduction in the final volume for disposal, assuming that evaporative or filtering type processes will be used to separate or concentrate the radioactive elements from the bulk liquid. In other cases, the liquid waste might be conditioned in situ (e.g. by cementation),

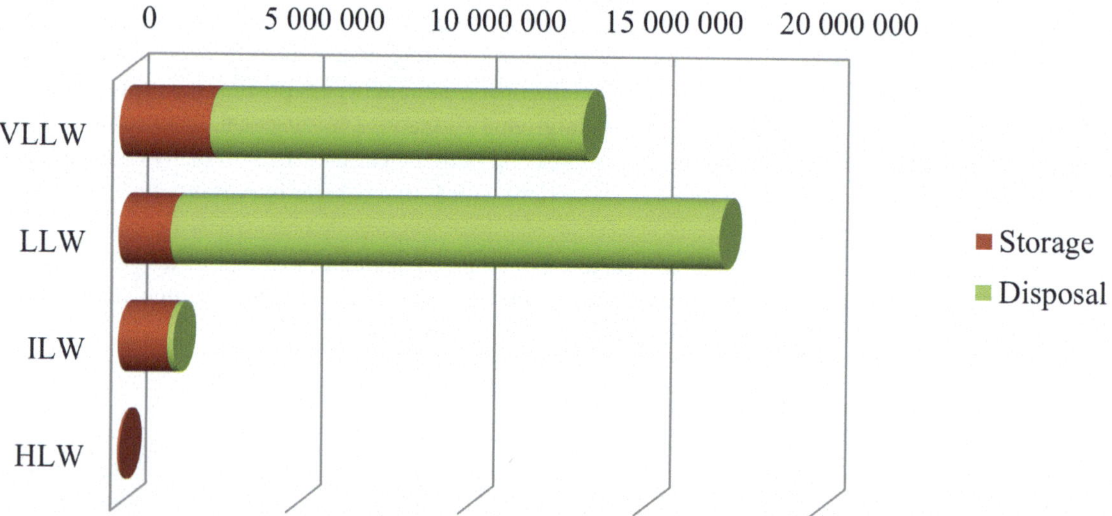


*FIG. 16. Summary of reported global solid radioactive waste inventories (m³). HLW storage volume is 54 000 m³.*

TABLE 8. REPORTED SOLID RADIOACTIVE WASTE IN STORAGE, AS OF 31 DECEMBER 2019

| Region | VLLW (m$^3$) | LLW (m$^3$) | ILW (m$^3$) | HLW (m$^3$) |
|---|---|---|---|---|
| Africa | 14 000 | 27 000 | 1000 | 0 |
| Americas | 2 309 000 | 471 000 | 54 000 | 7000 |
| Asia | 2000 | 248 000 | 106 000 | 1000 |
| Europe | 301 000 | 733 000 | 1 249 000 | 22 000 |
| Oceania | 0 | 5000 | 2000 | 0 |
| Global total | 2 626 000 | 1 484 000 | 1 412 000 | 30 000 |
| Joint Convention Contracting Parties | 2 624 000 | 1 438 000 | 1 412 000 | 30 000 |
| EU Member States | 292 000 | 392 000 | 321 000 | 5000 |
| OECD/NEA Members | 2 542 000 | 1 100 000 | 1 401 000 | 30 000 |

TABLE 9. REPORTED SOLID RADIOACTIVE WASTE IN DISPOSAL, AS OF 31 DECEMBER 2019

| Region | VLLW (m$^3$) | LLW (m$^3$) | ILW (m$^3$) | HLW (m$^3$) |
|---|---|---|---|---|
| Africa | 0 | 31 000 | 0 | 0 |
| Americas | 9 832 000 | 12 521 000 | 91 000 | 0 |
| Asia | 2 000 | 111 000 | 0 | 0 |
| Europe | 453 000 | 3 003 000 | 88 000 | 0 |
| Oceania | 432 000 | 24 000 | 0 | 0 |
| Global total | 10 719 000 | 15 690 000 | 179 000 | 0 |
| Joint Convention Contracting Parties | 10 719 000 | 15 660 000 | 179 000 | 0 |
| EU Member States | 453 000 | 1 077 000 | 14 000 | 0 |
| OECD/NEA Members | 10 717 000 | 14 697 000 | 105 000 | 0 |

which can result in a volume increase. In addition, the classification of the final waste package to VLLW, LLW, ILW or HLW will also depend on the length of time (i.e. radioactive decay time) that the waste has been in storage, the efficiency of the treatment process and the degree of volume reduction achieved.

## 6.5. FUTURE FORECASTS

To plan adequately for the long term management of radioactive waste, it is necessary to forecast the waste quantities and qualities expected in the future. This is a task that requires careful attention, especially for States with numerous and diverse activities that result in the creation of radioactive waste and with many different organizations involved in producing and managing it. For many States, radioactive waste generation is closely related to electricity production from nuclear power plants, and thus future forecasts are closely related to predictions for the future use of nuclear power.

These countries, which were nuclear pioneers and were involved in the initial development of nuclear power, may also need to deal with significant quantities of waste associated with the decommissioning and remediation of these early facilities and sites. Usually, the amounts of waste arisings from operation and decommissioning of the older facilities are much higher than similar waste arisings from present day nuclear power generation industry. Higher waste arisings have been noted on several occasions, including in China, France, the Russian Federation, the UK and the USA.

It is nevertheless important to make predictions regarding future waste arisings and to update them at regular intervals. This is important for the planning of facilities needed for storage, treatment and disposal and for establishing adequate funding for future waste management. It should also be recognized that precise numbers are not required to establish a reasonable basis for predicting future needs, as long as the inherent uncertainty in the quantity of future waste arisings is acknowledged. In due course, more precise data on waste volumes and radioactivity levels will be needed at the time of licensing of such facilities.

Defining the planning assumptions also requires proper consideration and attention. A major aspect to consider is the time frame for the forecast: the longer the forecast, the less accurate it will be. The amounts and composition of waste from different practices will vary according to:

— How facilities operate;
— The industrial and technical processes that generate the waste;
— The policies and regulations governing the industry, technology and waste management;
— The economics of different areas of the waste management cycle and waste management philosophies.

In developing forecasts of future waste quantities, the following considerations need to be addressed:

(a) Planning scenarios for future generation of electricity from nuclear energy (e.g. high, low and best estimates and constraints);

(b) Future operating strategies for the facility, such as the merits of waste minimization versus cost minimization;

(c) Timing of activities that impact on waste arisings, such as short term versus delayed strategy for remediation actions and solutions;

(d) Practical tools for creating forecasts (e.g. historical data, data from similar countries and facilities, engineering estimates);

(e) Modelling and process mapping (e.g. understanding the route through different waste management systems and the impacts of this on forecasts);

(f) Accuracy requirements (e.g. significance of impact on the waste owner or WMO if estimates are significantly too high or low).

In general, a 'bottom up' forecasting approach provides the greatest accuracy and facilitates the highest degree of flexibility. Using this approach, individual waste streams or categories are estimated at a facility level and then aggregated with other waste streams to produce an overall estimate. Initial estimates for planning purposes can also be derived by extrapolating past history, taking into account the number of facilities operated and their lifetimes; this latter approach is often simpler and more pragmatic. Estimates of future waste arisings should include all activities and life cycle phases of a facility that result in the production of radioactive waste, such as operation, maintenance, refurbishment and decommissioning.

Compared to LLW and ILW, spent fuel arisings are somewhat easier to forecast. The amount of spent fuel produced is broadly proportional to the amount of energy extracted from it. With knowledge of the number and type of reactors and their historical fuel burnup levels, a reasonable forecast can be made of future spent fuel arisings for the remaining lifetime of a reactor.

In some countries, forecasting of waste volumes has been performed over many years and forecasts are published regularly. Other States have only recently begun to undertake rigorous forecasting activities. Consequently, not all States have reported forecasts in their National Profiles. It should be noted that the Joint Convention [3] does not require the reporting of forecasts, only values of presently stored and disposed waste. However, the Euratom Waste Directive [5] requires that forecasts of the future generation of radioactive waste be reported.

# 7. SYNTHESIS OF ACHIEVEMENTS AND OVERARCHING ISSUES

The first publication in this Status and Trends series [1] was more focused on collecting information on spent fuel and radioactive waste management practices and inventories, while the second publication provides an additional analysis of global trends. The first revision [2] also included discussion of significant achievements; discussion and analysis of current and emerging issues; and reports on progress in addressing previously identified issues. This current revision provides an update on that analysis, so this section focuses on a discussion of the achievements and general issues related to spent fuel (whether it has been declared to be a waste or not) and to the various radioactive waste categories.

## 7.1. MANAGEMENT OF SPENT FUEL AND HIGH LEVEL WASTE

Information concerning the implementation of the back end of the nuclear fuel cycle is presented in Section 5. Several countries have made decisions on the policy to recycle or to dispose of the spent fuel, but the options are being kept open or the topic is still under discussion in many countries. There are different factors that influence Member States in choosing an open or closed cycle.

A theoretical study by the NEA [49] looked at the economics of the back end of the nuclear fuel cycle, comparing higher level idealized open and closed cycle systems. The main factors that influence Member States in choosing their spent fuel management strategy could be divided into following groups:

— Political/social;
— Strategic;
— Economic;
— Environmental impact;
— Non-proliferation/security considerations.

The spent fuel management policy is defined at the national level, even if the nuclear operators are privately owned companies. The main factor leading to the choice of a spent fuel management policy in the frame of a long term vision is usually the nuclear power policy and strategy. Several countries have acknowledged that policy may change based on developments in technology, economics, public perception and environmental considerations. Flexibility in a spent fuel management strategy and/or nuclear waste management strategy may be an additional desired factor in decision making processes, especially in the case of States that are newcomers to nuclear power.

In general, countries with very large nuclear programmes, such as China, France, Japan, the Russian Federation, the UK and the USA, have implemented their spent fuel management strategy with a long term view based on country specific considerations. The UK and the USA are following an open cycle policy, while others have chosen the closed cycle with the long term view of implementing Generation IV reactors. Delays in such implementation could occur and may raise some concerns related to the management of the recovered materials if not recycled in the Generation III reactor fleet. Countries choosing the open cycle generally have no immediate interest or purpose in using the uranium and plutonium recovered from reprocessing of the spent fuel.

The approaches of small to medium size nuclear power countries are diverse. Most of them have opted for the open cycle route considering both non-proliferation and economic aspects. This is the case, for instance, for Finland and Sweden. Others, such as the Kingdom of the Netherlands, have chosen the closed cycle route with full recycling of the valuable materials, with the drivers being the cost certainty and the environmental impact of storing/disposing of HLW instead of spent fuel. This also provides additional flexibility for possible shared disposal solutions in the future. Looking at possible future flexibility, other countries, such as the Czech Republic, Hungary, Slovenia and South Africa, have chosen the open cycle as a reference but continue to examine alternatives.

Some countries, such as Argentina, Belgium, Brazil and Ukraine, have chosen to implement storage of spent fuel whilst maintaining the flexibility to decide on an open cycle or closed cycle policy in the future. However, it should be kept in mind that storage is just an intermediate step in spent fuel management and cannot be considered as an end point.

Although the rate at which spent fuel was reprocessed was more or less constant, the recent trend has been for fewer States to send their spent fuel for reprocessing overseas, with the amount of spent fuel in long term storage increasing. Reprocessing capacities have been reducing because only a few new plants are under construction and several reprocessing plants have been closed or are nearing the end of their lifetime. The development of the nuclear programmes in China, India and the Russian Federation could change this trend. However, recycling of spent fuel continues to play an important role and there is a focus on developing multirecycling technologies to be applied using thermal reactors that can provide a sustainable solution for the transitioning period from once through recycling (currently implemented industrial cycle) to a fully closed fuel cycle with fast reactors. There are examples of countries considering innovative recycling technologies to reduce the burden of generated wastes and the footprint of disposal facilities by recycling long lived products for medical applications. Unfortunately, implementation of fast reactor based closed fuel cycles is being delayed in many countries, although research and development activities on large reactors and associated fuel cycles continue in some of them [50–52].

Long storage periods introduce several challenges in terms of technical aspects (e.g. changes of technologies, ageing management of facilities, etc.), licensing (changes of regulation), organization (changes in the nuclear industry) and funding (accuracy of costs in the long term and availability of funds) and the challenge is thus to ensure the long term safety and integrity of the storage facilities and the spent fuel/HLW for many decades to come. A number of national strategies reflect the need to make sufficient spent fuel storage capacity available to bridge the gap between the generation of spent fuel and the foreseen commissioning and operation of deep geological disposal facilities. There is evidence of greater attention being paid to the impacts of the fuel cycle on disposal, and vice versa, especially in view of some uncertainty regarding the requirements and acceptance criteria of the disposal facilities. There is an urgent need to work to understand and optimize the whole back end and to actively implement these strategies on the ground [50]. However, such challenges are properly managed, and countries have decades of experience with spent fuel storage, both in wet and dry storage facilities. There are also active programmes in place to monitor the condition of the spent fuel and its storage environment to ensure that it can be safely stored for the required length of time. This already motivates improvement of existing storage facilities, for example in Finland. Requirements and issues related to spent fuel storage are discussed in further detail in Ref. [11].

As the requirements for storage capacity increase, new storage is built outside of the reactor buildings. Owing to the need for longer storage, there have been successful implementations of storage facilities that are planned, built and operated either in the vicinity of the reactor building or as a centralized facility in the country [11]. Most of these are facilities for dry storage, but some pool facilities are also in operation. Depending on the strategy for spent fuel management, away from reactor (AFR) facilities using wet or dry storage technologies have been licensed and built. There are AFR on-site facilities, for example, in Belgium, Canada, Hungary, Spain and the USA, and centralized off-site AFR facilities in Germany, the Kingdom of the Netherlands, Sweden and Switzerland. Currently, approximately 80% of the AFR facilities are based on dry technologies and this is mainly owing to their modular and passive nature. To be economic, a wet storage pool generally needs to be large and hence has fixed capacity. Dry storage facilities, especially of the cask type, can be built to any size scale and can be expanded incrementally over time. This means that not all of the cost is required up front, unlike in a fixed capacity wet pool.

Some recent progress and achievements in development of the deep geological disposal facilities include:

— In Belgium, the legal process for adoption of geological disposal was launched in 2022.
— In Canada, the site selection process has progressed from 22 to 2 possible communities, with a single preferred site to be identified in 2023.
— In China, the construction of the Beishan Underground Research Laboratory in Gansu Province to support research into geological disposal of HLW in crystalline host rock started 2021 and is expected to be completed in 2027. The aim is to construct a geological repository by approximately 2050.
— In Finland, the operational licence application for an encapsulation plant and geological disposal facility in Onkalo site was submitted to the Government in 2022. This is a first of a kind in the frame of a DGR. The construction of underground access tunnels at the ONKALO facility in Finland started in December 2016 and the construction of the encapsulation plant started in 2019. The first construction phase for the deposition tunnels is complete (five deposition tunnels excavated).
— In France, the detailed concept for an HLW and ILW disposal project was achieved in 2020 and the environmental impact assessment was issued in 2021. The project was recognized as a final disposal solution for the most radioactive waste produced in France in 2022, and it is expected that the construction licence for HLW deep geological disposal will be submitted in 2022.
— In Hungary, there is ongoing siting survey framework programme for a national DGR, with siting in the Mecsek region.
— In Lithuania, a general DGR concept will be developed by 2023; the selection criteria for siting are in preparation and the site should be selected by 2047. It is planned that the facility will be operational by 2068.
— In Sweden, in January 2022 the Government licensed an encapsulation plant and disposal facility for spent fuel.
— In Switzerland, the proposed site was announced in 2022. A general licence application is expected to be submitted around 2024, and the government decision is expected by 2029. There is a possible national referendum on the disposal facility. The plan is to build a pilot disposal facility, which will be closed after testing, and the real disposal facility will be built in neighbouring rock.
— In the UK, the process to identify a suitable site for DGF in England or Wales is underway. Four voluntary communities are engaged through community partnerships. Non-intrusive seismic investigations to gain better understanding of the geology off the coast of Cumbria started in 2022.

The site selection approaches taken by the Member States can be different — some are taking a technical siting approach, while others are taking a voluntary approach or a combination of technical and voluntary. In all cases, the importance of public acceptance has been recognized and all of the programmes include extensive public consultations at various points in the process. Thus, the DGR time schedules have been revised, taking into account the realistic timescales for the siting activities, public acceptance, technical implementation and regulatory activities owing to the challenges caused by a lack of experience in licensing such facilities, etc. Several countries have had to restart their site selection process for a DGR, having not been successful in gaining public support. Proposed timelines for the operations of DGRs varies from country to country, but to date the estimated earliest and latest years for opening DGRs are 2024 and 2160, respectively.

It is obvious that for some countries with a limited inventory, in particular for countries that are only operating research reactors, the cost of development of disposal facilities may be very high compared to the benefits from their operation. At present, several countries return the research reactor fuel to the country of origin of the fuel if possible, and thus it does not require disposal in the country where it has been used. The Russian Federation and the USA have had agreements to take back the research reactor fuel from the countries where it was used. There are some countries that have decided to reprocess research reactor fuel, such as Australia and Belgium, for instance, which have made the choice to reprocess in France. The international experience accumulated from research reactor fuel take back programmes for high enriched uranium has been collected and presented in the IAEA Nuclear Energy Series No. NW-T-1.11, Available Reprocessing and Recycling Services for Research Reactor Spent Nuclear Fuel [42].

When most of the storage systems were put into operation, they were designed to last between 20 and 50 years. Over the last six decades of wet storage, and four of dry storage, good performance has been reported and valuable operational experience gained. As these systems reach the end of their originally intended lifespans, work is underway to develop additional monitoring and inspection techniques to support ongoing safe spent fuel storage, as well as the licensing or relicensing of these activities.

Ongoing research and development (R&D) is an important and integral part of most radioactive waste and spent fuel management programmes. Nuclear power countries have extensive programmes for research and development in spent fuel and radioactive waste management. R&D is carried out by a variety of entities, including facility operators, regulators and technical support organizations, as well as by independent organization, such as universities and research institutes. The specific goals of the R&D can vary from basic science fundamentals to applied research for developing specific technical solutions. Some of the current and proposed research is centred on the development and demonstration of technical equipment required for repository construction and operation (e.g. construction methods, waste handling/emplacement equipment, tunnel sealing, etc.). Collaborative work is also ongoing in areas such as ageing management, predisposal management, spent fuel characterization, high burnup issues, partitioning and transmutation, which could have an impact on the development of advanced fuel cycles, as well as the types and quantities of waste to be disposed of. The development and potential deployment of new reactor types, such as small modular reactors, will also affect the need for and the type of fuel cycle and waste management facilities in countries wishing to deploy such technology.

There is ongoing R&D to reduce the volume and potential hazard of HLW, for example separation and transmutation/conversion of actinides would help to reduce hazards related to the management of spent fuel, especially long term storage or disposal. For instance, an advanced fuel cycle that includes partitioning and transmutation in addition to the reuse of uranium/plutonium (e.g. in Generation IV fast reactors or accelerator driven systems) could provide benefits for geological disposal by reducing the radiotoxicity and thermal output of the final waste inventory, thereby positively impacting the required footprint of such a repository. However, it should be noted that, since such a fuel cycle is still in the R&D phase, the actual benefits for the fuel cycle in general and geological disposal in particular are difficult to estimate. There is R&D related to different aspects of small modular reactors and Generation IV reactors. For example, Belgium has approved the Myrrha (Multipurpose Hybrid Research Reactor for High-tech Applications) project and construction is expected to begin in 2026.

The safety case is the collection of all arguments that contribute to demonstrating the safety of the facility and includes calculation/modelling (quantitative safety assessment), especially as, because of the very long time frames covered by the safety cases for radioactive waste management and spent fuel management, the safety cases are generally based on mathematical modelling and simulation and/or by comparison to suitable natural analogues, rather than by direct observation of repository performance. To develop, test and calibrate the models, research and development is generally required to establish model parameters and boundary conditions (e.g. diffusion rates, radionuclide transport processes and corrosion mechanisms under repository conditions; design parameters, such as rock strength and other properties;

etc.). In the shorter term, R&D may also be required to demonstrate the various technologies required to construct and operate a disposal facility; create optimized waste forms; develop technical specifications and waste acceptance criteria, etc. The exact requirements for R&D will vary by country and by waste type and the planned management routes. Often, a certain amount of initial R&D is required in order to be able to decide the appropriate management route.

As stated previously, the length of the storage period could be many decades. There can be regulatory provisions and technical measures defined, which should be followed if it is necessary to ensure continued safe storage over the longer term. A future challenge is related to handling of the spent fuel and its integrity after several decades of storage, especially in dry storage conditions where it is not possible to directly observe the fuel because it is sealed/bolted in casks or canisters.

Sharing of knowledge and experience needs to be encouraged among the countries under the auspices of international organizations such as the EC, the IAEA, the NEA and the WNA. Some examples are provided below.

The IAEA has conducted and is conducting a series of activities and Coordinated Research Projects on different topics related to spent fuel management, such as:

— Spent fuel performance (SPAR-I to IV series);
— Spent fuel research and assessment (SFERA);
— Ageing management programmes (AMPs) for dry storage systems;
— Performance assessment of storage systems for extended durations (PASSED);
— Challenges, gaps and opportunities for managing spent fuel from small modular reactors.

The behaviour and integrity of spent fuel and cladding materials are paramount, since they are the first barrier for radioactivity containment during long term storage and subsequent fuel handling operations. Summaries of the results concerning this were published in successive IAEA technical publications [51, 52], including an IAEA technical document summarizing the most relevant findings in the context of today's implementation of the technology [53].

In recent decades, national and international R&D programmes have been conducted on the development of advanced sustainable nuclear fuel cycles associated with Generation IV reactors in order to improve the utilization of uranium resources, maximize energy production, minimize waste generation, improve safety and limit proliferation risks. Advanced fuel cycles are devoted to recycling most of the long lived minor actinides, minimizing not only the volume of the high level waste to be finally disposed of, but also its radiotoxicity as well as the decay heat, thereby lowering the burden of the waste and the 'repository footprint'.

The NEA has several working groups[6] and some to mention in the relation to the safe management of spent fuel and radioactive waste are:

— The Integration Group for the Safety Case (IGSC) on deep geological disposal, particularly for long lived and high level radioactive waste. The mission of the IGSC is to assist member countries to develop effective safety cases supported by a robust scientific–technical basis. There is also an ad hoc group on transfer and return of gained experiences on safety cases for disposal facilities (TARGES).
— The Expert Group on Building Constructive Dialogues Between Regulators and Implementers in Developing Disposal Solutions for Radioactive Waste (RIDD) has compiled, inter alia, a report on Building Constructive Dialogues Between Regulators and Implementers in the Pre-Licensing Phase of a Deep Geological Repository [54] and works on the establishment of a generic roadmap towards licensing of DGR facilities[7].

---

— The Horonobe International Project aims to develop and demonstrate advanced technologies for use in accordance with DGR design, operation and closure, as well as realistic safety assessment, which are recognized as common international challenges.

The WNA has created a Sustainable Used Fuel Management Working Group, with the main objective of gathering the views of the nuclear industry and stakeholders (including newcomers) on the back end of the fuel cycle. The working group considers how the industry can best respond to these needs, as well as explaining how effective spent fuel management contributes to the sustainability of nuclear energy and supports the development and implementation of it. The Working Group on Waste Management and Decommissioning is covering a wide range of topics and will also work on incorporating waste management and decommissioning lessons into small modular reactor design and propagating the circular economy in the nuclear industry through material and waste management.

In addition to their formal research programmes, the IAEA, NEA and WNA all host and support peer to peer networks and expert working groups on a range of topics related to radioactive waste management, spent fuel, fuel cycles and decommissioning. These networks and working groups facilitate the exchange of information and experience among their members.


## 7.2. MANAGEMENT OF RADIOACTIVE WASTE

Globally, the volumes of ILW are small compared to LLW and VLLW, typically less than 5% of the total. Many industrial scale methods exist for safe processing, packaging and storage of ILW. Many countries (e.g. Belgium, the Czech Republic, France, Japan, Switzerland and the UK) plan to develop an underground disposal facility for ILW co-located with HLW. There are some new combined ILW and LLW disposal facilities in operation. Additionally, one is under construction (Konrad in Germany is licensed and under construction; commissioning is scheduled for 2027) and two are in the regulatory approvals process (Cigéo in France, SFL in Sweden, which is planned for the late 2030s). In Canada, the proposed facility for disposal of LLW and ILW at the Bruce site was cancelled after a non-supportive vote from local indigenous communities.

Special challenges are connected to some categories with larger volumes (e.g. graphite from gas cooled reactors and radium bearing waste from earlier radium production). Although the radioactivity levels in some of the waste can be relatively low, it mainly comprises long lived radionuclides, such as $^{14}$C and $^{226}$Ra, together with the decay products, and is therefore managed as ILW. The issue is to develop a solution that is proportionate to the actual hazards of these wastes, which are less active than some other ILW waste.

Historically, disposal solutions have been developed first for low level waste. The classification and the safety criteria, in particular for the long term, have been established gradually and they have mainly addressed operational waste. There are a number of countries with disposal facilities for LLW but owing to the planned decommissioning of their nuclear installations, there might be a need for additional capacity in the near future.

The amount and characteristics of waste and other materials to be generated by a given decommissioning project are related to several factors, including the reactor type and design, unit history, decommissioning strategy, safety and environmental regulations and radioactive waste management routes. However, typically decommissioning activities imply the generation of large volumes of materials that need to be properly managed. Only a small fraction of that waste will generally be classified as radioactive waste. The amount to be declared as radioactive waste will vary by country, according to their laws, regulations, practices and available waste management infrastructure. For example, it is estimated that approximately 104 000 t of materials will be managed throughout the duration of the José Cabrera nuclear power plant dismantling project in Spain. Approximately 4% of them will be classified as radioactive waste.

For those materials from decommissioning that are not considered as radioactive waste, it is common to follow reuse and recycling approaches. The main fraction of residual materials generated during these activities will normally become 'conventional' wastes that will be managed through standard industrial waste management routes and outside of nuclear regulatory control. Specific attention is required for those non-radioactive wastes causing toxic or chemical hazards (e.g. heavy metals, asbestos).

Radioactive decommissioning waste is mostly similar to radioactive operational waste in terms of radiochemical hazard and risk. Similar approaches and technologies are used for its treatment and conditioning. However, dedicated waste streams may be generated that require specific consideration.

Most disposal facilities for LLW are surface or near surface facilities, including relatively shallow depth underground caverns. Surface or near surface facilities can be found, for example, in France, Romania and Spain. However, in several countries, such as Germany, Hungary, Sweden and Switzerland, the choice has been made to dispose of this waste in deeper rock formations.

The development of disposal facilities dedicated to this waste is continuing to progress globally, with some examples given below:

— In Brazil, the selection process, as well as the conceptual design, for a disposal facility (National Center for Nuclear and Environmental Technology (CENTENA)), is in the final stages.
— In China, Longhe near surface disposal site (within a region of the Gobi Desert) has been confirmed and approved as a centralized disposal site for LLW from nuclear power plants and the facility is now operational. Extensions have been implemented at existing low and intermediate level waste (LILW) disposal sites in China (Northwest and Feifengshan).
— In Bulgaria, a national near surface disposal facility for LLW is at the construction stage.
— In Italy, there is a search for a site for LLW and ILW disposal; the eligible areas were published in 2021. This process will assess interest from local communities at one or more sites from a list of identified candidates.
— In the Republic of Korea, an engineered shallow land disposal facility at Gyenongju is under regulatory review for construction and operation.
— In Norway, there is a LILW storage and disposal facility at Himdalen, with plans for assessing future capacity.
— In the Russian Federation, the first section of a near surface disposal facility for solid class 3 and 4 radioactive waste at the UECC site, Novouralsk, is in operation, and the second stage under construction. The licensing process for near surface disposal facilities at Chelyabinsk and Tomsk has been started. The near surface disposal facility in Sergeiev Posad near Moscow is at the planning stage.
— In Slovenia, it is planned to start construction of a LILW repository in 2022.
— In Spain, the El Cabril disposal facility submitted an application for capacity extension in 2022.
— In Sweden, the extension of the SFR repository for short lived LLW and ILW was approved in 2021. The main reason for the extension was the need to accommodate decommissioning waste.
— In the United Arab Emirates an application for a near surface waste disposal facility was made in 2022.
— In the UK, LLWR is for the disposal of LLW, while landfill sites are also available for the least hazardous LLW.

Several new disposal facilities have been expanded. One example is a facility in Hungary (Bátaapáti), where a second disposal chamber has been completed. There are also several site selection processes being undertaken internationally; for example, in Australia, Malaysia and Pakistan.

Existing disposal facilities are considered to be valuable assets that are difficult to replace and waste minimization is an important means to expand their lifetimes: for the LLWR (Low Level Waste Repository) facility in the UK the lifetime has been expanded through a successful waste diversion programme, with up to 85% of new waste being diverted, resulting in the diversion of some 50 000 m$^3$ during the period 2008–2016. The operator promoted decontamination and recycling options as well as diversion to VLLW landfills. This is a concern shared by all disposal facility operators. A waste minimization approach is also being applied in other countries using the definition of VLLW as a subclass that allows them to develop dedicated management and disposal facilities for such wastes (e.g. France, Spain).

Waste minimization is a challenge as decommissioning waste streams are increasing. To face this issue, an expansion of the SFR facility in Sweden to accommodate decommissioning waste is at the licensing stage. In the Republic of Korea, optimization is planned through expansion of the existing silo type disposal facility at Wolsong with the addition of a near surface facility for LLW.

The construction of a dedicated disposal facility for LLW in countries with small quantities of waste may be difficult because of the relatively high initial fixed cost to site, design, license and construct a repository. Alternative solutions, such as borehole disposal, may be more practical and cost effective to implement.

According to GSG-1 [4], very low level waste (VLLW) is a waste "that does not need a high level of containment and isolation and, therefore, is suitable for disposal in near surface landfill type facilities with limited regulatory control." Some States consider VLLW as a subclass of LLW and dispose of the two in different areas of the same facility (e.g. Spain, the USA). Others have created separate facilities for the two types (e.g. France, Lithuania, Sweden). Others still have licensed appropriately permitted landfill sites to accommodate VLLW (e.g. the Kingdom of the Netherlands, the UK). The implementation of disposal facilities dedicated to very low level waste is the result of a desire for optimization and to reserve LLW disposal facilities for wastes that truly require the higher level of engineered barriers and isolation provided by them.

The relevance of this category will increase during dismantling operations of nuclear facilities: large volumes of VLLW are expected in the future as a result of decommissioning and dismantling programmes for nuclear power plants (see Section 3.2.1). Preparedness to accommodate large quantities of waste for disposal over a fairly short period of time will be important — particularly in countries that foresee an accelerated decommissioning programme for nuclear power plants. In addition, much of the waste produced as a result of dismantling a nuclear facility is different from normal operational waste, as it has a higher proportion of minimally contaminated metals (e.g. equipment) and building rubble. Cost effective disposal solutions that still provide adequate safety for this class of waste have been and are being developed in various countries.

There are already several examples of activities to increase disposal capacities for VLLW. Often the site or planned area will not be changed, but with optimization of disposal the capacity will be increased. For example, in France it is planned to submit a licence application to increase the disposal capacity for a VLLW disposal facility in 2022 and Hungary introduced the VLLW subclass for waste.

However, despite this relevance, it should not be forgotten that, as for LLW, disposal facilities for VLLW are rare assets and their availability has to be preserved for as long as possible. There is a risk of early saturation of disposal capacities in some countries, such as France. Therefore, countries are also focusing on alternative solutions [55–56]:

— Waste minimization at the source: characterization may be a significant challenge as the lower the activity, the longer and the more difficult activity assessment is. Reliable industrial assessment is required. Generally, characterization of VLLW is also connected to free release issues, to segregate waste to be managed in this specific radioactive route from exempted waste. These challenges are approached by related research and development, supporting operational experience.
— In several countries, it is possible to avoid production of radioactive waste to some extent by implementation of recycling of radioactive material. Again, adequate characterization is a major consideration and also there have to be clear regulatory processes. It has been successful in some countries, such as Spain, Sweden and the UK, illustrating the impact of national policy and regulatory frameworks; however there are also countries where recycling of radioactive materials is not permitted.

Management of large components constitutes a dedicated point of interest as this activity is usually particular to decommissioning. Different approaches have been taken to date in response to the peculiarities of individual projects and associated management routes. For example, in some countries, such as the USA, large components can be disposed of in one piece, perhaps after stabilization by filling them with grout. In other countries, such as Germany, the prevailing practice is to cut the large components into smaller segments that will fit inside a 'standard' waste package.

Most operating radioactive waste disposal facilities are for a given class or type of waste and will accept waste from any source as long as it meets the acceptance criteria for the facility (e.g. it can take waste from either operation or decommissioning of a nuclear facility). However, there are several cases in which such facilities are licensed for operational waste but not for decommissioning waste (e.g. Finland, Japan, Sweden) and others (e.g. Spain) in which technical modifications have been needed to enable a more efficient use of available disposal capacity (licensing of 'larger decommissioning packages').

The disposal of DSRSs is still a challenge in most countries, and new concepts are currently being developed. Most Member States still only have arrangements in place for storage, sometimes as a long term solution. There are several ongoing international initiatives to safely manage DSRSs and one of these involves the construction of the first borehole for disposal of DSRSs. The final disposal solution for DSRSs (e.g. borehole, near surface, geological, etc.) has to be optimized, taking into account social, economic, technical and safety aspects. The Malaysian Nuclear Agency received a licence application for the borehole disposal of its inventory of disused sealed radioactive sources in 2019. In 2022, the IAEA contracted a drilling company to construct the borehole with implementation through the Malaysian Nuclear Agency and an IAEA team assembled a mobile facility on-site that will be used for conditioning the DSRSs into the disposal packages. The actual disposal operations are expected to begin at the end of 2025 or the beginning of 2026.

Some examples of current R&D in the field of long term radioactive waste management include:

(a) Country specific R&D: very active R&D programmes in several countries (e.g. Canada, France, Sweden, Switzerland, the UK, the USA):
   (i) These normally focus on demonstrations of equipment and technology, improving the science and understanding of issues related to long term safety (e.g. corrosion of containers, behaviour of engineered barrier materials under repository conditions, geosphere characterization, etc.).
   (ii) Some are projects undertaken by individual countries and some are collaborations between multiple countries. There is a move towards collaboration because the costs to develop, operate and maintain facilities are high. Most of the work is common to a number of programmes anyway, so there is a cost benefit in collaborating and leveraging funding.
(b) The framework programme in the management and disposal of radioactive waste under the Euratom Horizon 2020 programme.
(c) The NEA NI2050 programme on waste and decommissioning — waste related issues are recognized as being important for sustaining the nuclear power industry into the future.
(d) IAEA coordinated projects — these cover a wide range of topics, from the development of a standardized framework for the borehole disposal of DSRSs and small amounts of low and intermediate level waste to spent fuel storage.


7.3. OVERARCHING ISSUES

The radioactive waste and spent fuel management programmes in the different Member States have unique characteristics and aspects related to the peculiarities of the programmes, such as the size of their power programme or their degree of maturity. Likewise, it is also possible to identify aspects and issues of common interest for many of these national programmes (overarching issues) that invite the development of joint actions to overcome them. This aspect has been developed by the subsequent Joint

Convention review meetings; at the last meeting in June 2022 the following eight proposals among those in the working sessions were highlighted:

(a) Competence and staffing linked to the timetable for spent fuel management and radioactive waste management programmes. Effective and efficient implementation of policies and strategies for spent fuel and radioactive waste management depends on the availability of suitably qualified and experienced personnel across all organizations involved in the management of spent fuel and radioactive waste. The Contracting Parties discussed and emphasized the importance of knowledge management while recognizing the extended timelines involved in safe storage of spent fuel and radioactive waste and development of disposal facilities.

(b) Inclusive public engagement on radioactive waste management and on spent fuel management programmes. Inclusive, open and transparent engagement with the public and understanding the role of all organizations involved in the process was emphasized as a key factor in enhancing public trust. The Contracting Parties discussed and highlighted the value of undertaking an inclusive approach by not only focusing on disseminating information but also engaging with and listening to the public and relevant stakeholders in discussions.

(c) Funding of long term projects. The discussions in several of the Country Groups highlighted challenges with respect to securing funding for radioactive waste management and spent fuel management, noting the long timescales and the slow realization of disposal facilities. The Contracting Parties during the discussions recognized that the provision of funding is required under the Joint Convention, Article 22. Thus, there is already an obligation for Contracting Parties to report on this issue.

(d) Management of radioactive waste and spent fuel from new technology applications as well as planned new projects using existing technologies. Contracting Parties underlined the need for proactive work to conduct research and to develop strategies for the management of novel spent fuel and radioactive waste arising from new nuclear power plants (NPPs) or new technologies. However, some Contracting Parties noted that this topic was not applicable to all Contracting Parties. It was suggested that this topic be considered for the Topical Session at the Eighth Review Meeting, subject to the agreement of Contracting Parties at the Organizational Meeting of the Eighth Review Meeting.

(e) Legacy wastes linked to decommissioning and remediation projects. Several Contracting Parties reported progress in the decommissioning of legacy facilities and the remediation of legacy sites. Some Contracting Parties identified the need for the establishment of a national strategy for dealing with legacy sites, including the need to build new facilities for the safe management of the waste arising. Some Contracting Parties have reported that the remediation of uranium mines has remained a major technical and financial challenge.

(f) Ageing management of packages and facilities for radioactive waste and spent fuel, considering extended storage periods. This issue was linked to the absence of the timely availability of disposal facilities as well as the fact that some Contracting Parties have national policies regarding spent fuel being considered as an asset, resulting in extended periods of storage. It was further noted that this topic is relevant to the intergenerational equity that is covered in the text of the Joint Convention.

(g) Realization of disposal facilities. The Contracting Parties considered this issue linked to the one above. While near surface disposal facilities are in place in several Contracting Parties, few geological disposal facilities are under consideration or implementation by some Contracting Parties. In this context, it was underlined that decisions and actions for the realization of disposal plans should be thoroughly considered and driven by safety objectives.

(h) Long term management of disused sealed sources, including sustainable options for regional as well as multinational solutions. The Contracting Parties considered that the management of orphan and disused sealed sources remains an overarching issue since the Fifth Review Meeting. The availability of disposal routes and uncertainties in the availability of transborder solutions for disused sources was highlighted in the Country Group discussions. It was noted that many Contracting Parties consider disused sealed sources as waste, while others recycle and reuse them. Nevertheless, the safety of long term management of disused sealed sources was identified as a challenging issue, which would be worthy of increased attention at the Eighth Review Meeting.

# 8. ANALYSIS OF TRENDS

This is the third publication in this series and its main purpose is to highlight both the status of and noted trends in spent fuel and radioactive waste management. The general trends are presented based on information provided in the National Profiles, in discussions during the Seventh Review Meeting of the Contracting Parties to the Joint Convention (held in 2022) [8], and in the Report from the Commission to the Council and the European Parliament on progress in the implementation of Council Directive 2011/70/ EURATOM and an inventory of radioactive waste and spent fuel present in the Community's territory and the future prospects [6]. Inputs from the different international working groups (e.g. Technical Working Groups on Radioactive Waste Management and Technologies and on Nuclear Fuel Cycle Options and Spent Fuel Management) were used as well.

This section on trends is divided into three subsections. First, general trends are considered in Section 8.1, and then trends related to radioactive waste management are discussed in Section 8.2, followed by trends related to spent fuel management in Section 8.3. This section outlines some generally recognized trends that have been identified since the publication of the first revision of this publication [2]. For convenience, they are grouped into a number of functional areas or themes.

## 8.1. GENERAL TRENDS

### 8.1.1. Policy and strategy

The policies and strategies for radioactive waste, spent fuel management and decommissioning are generally developed and set by government ministries or agencies. Defining long term aims in the policies usually helps to ensure successful and optimized implementation of strategies. Many countries have successfully established policies and strategies that have been stable over several decades and good progress has been made towards their implementation. An overview of defined aims can be found in Annex II.

Maintaining a stable policy and strategy over time can be a challenge for various reasons. For example, the decision made in Germany to phase out nuclear energy will also affect its spent fuel and radioactive waste management. In other cases, a national policy may have evolved around practices that have been established for decades, but these approaches or strategies might no longer be preferred given the circumstances of today. As the possible management options can also be changed, this means that revisiting the decisions made in the past is also important (e.g. the waste diversion programme Low Level Waste Repository in the UK). Spent fuel management is a long term commitment and the strategies adopted for managing the spent fuel produced by power reactors need to keep some flexibility to enable potential changes in policy decisions.

More and more countries are developing, or at least considering, integrated and holistic approaches to spent fuel and radioactive waste management. This has multiple benefits, including more effective use of resources, such as disposal space, and reduced overall costs. For nuclear newcomer countries, this is a very important consideration to ensure that spent fuel management and radioactive waste management systems are established in an optimal way right from the beginning. More attention and corresponding effort are put into ensuring consistency and safety in predisposal and disposal activities in radioactive waste management. Similar efforts can also be seen related to spent fuel storage and disposal activities.

Some overall tendencies can be seen in national decisions in spent fuel and radioactive waste management. Often countries opt for a single solution for management of the same type of waste or for all waste producers in a country. The siting process for disposal sites is usually undertaken by national waste management organizations, which are often independent of the waste producers/owners (Annex III). The siting process is often a broad community consent based engagement process, in which many factors are taken into account. There are safety related questions, but additionally the selection process includes estimation of the environmental impact and consideration of transportation issues, cost, etc.

### 8.1.2. Governance

Continuous improvement is a key component of radioactive waste and spent fuel management programmes. There are several review services available for the Member States and there is a clear tendency for these services to be used more than in the past. There are several reasons for this, including the fact that there is an increased recognition of the expert peer reviews organized by international organizations such as the IAEA and the NEA. In fact, such regular peer reviews are legally required for EU Member States under the Euratom Waste Directive [5].

The IAEA's Integrated Review Service for Radioactive Waste and Spent Fuel Management, Decommissioning and Remediation (ARTEMIS)[8] is an integrated expert peer review service for radioactive waste and spent fuel management, decommissioning and remediation programmes. This service is intended for use by facility operators and organizations responsible for radioactive waste management, as well as regulators, national policy makers and other decision makers. Between launching the service in 2016 and the end of April 2023, 23 review missions were conducted.

For newcomers in nuclear energy there is, for example, the IAEA Integrated Nuclear Infrastructure Review (INIR) [57], which is a holistic peer review to assist countries in assessing the status of their national infrastructure for the introduction of nuclear power. The review covers the comprehensive infrastructure required for developing a safe, secure and sustainable nuclear power programme. The topics covered also include the nuclear fuel cycle and radioactive waste management. Since then, INIR missions for the different phases of developing a nuclear power programme were hosted by the following Member States: Bangladesh, Belarus, Egypt, Ghana, Indonesia, Jordan, Kazakhstan, Kenya, Malaysia, Morocco, Niger, Nigeria, the Philippines, Poland, Saudi Arabia, South Africa, Sri Lanka, Sudan, Thailand, Türkiye, Uganda, the United Arab Emirates, Uzbekistan and Viet Nam.

Another key aspect is the growing interest of the public and other stakeholders in nuclear activities in general and radioactive waste and spent fuel management in particular, leading to an increased awareness of the role played by society in effective implementation of radioactive waste and spent fuel management programmes and projects, in particular related to disposal. This has led to many countries adopting a more open and transparent system for communicating with the public and other stakeholders, as well as to the development of more participatory regulatory review processes. The challenge is for the decision makers to balance technical considerations with sociopolitical and economic considerations.

### 8.1.3. Funding and financing aspects

The successful implementation of a radioactive waste and spent fuel management programme (and/or nuclear decommissioning programme) requires adequate and stable funding, often over many decades or longer. While most of the countries with nuclear power programmes have established dedicated funds to support these programmes and ensure that an undue burden is not passed on to future generations, the challenge is to ensure that the funds remain viable, secure and available over the long time periods required. The situation in each country will be different, and various mechanisms have been established in different countries to ensure the future adequacy of the fund. One important aspect common

---

[8] See   https://www.iaea.org/services/review-missions/integrated-review-service-for-radioactive-waste-and-spent-fuel-management-decommissioning-and-remediation-artemis

to many countries is the requirement for periodic review and assessment of the fund and its management. This includes a review of both the cost estimates and the investment strategy for the fund, as well as any economic parameters used in calculating the adequacy of the fund (e.g. inflation rates, interest rates, financial discount factors, taxation implications, etc.). It is clear that increased social and political interest will help to guarantee the adequacy and sufficiency of the funds available for the safe management of spent fuel and radioactive waste.

### 8.1.4. Regulatory framework

The Joint Convention [3] and several safety standards [21, 28] are publications that define the basis of international agreements for the safe management of spent fuel and radioactive waste. To secure the safety of the public and workers, a regulatory framework has been put in place in the majority of countries. There has to be collaborative effort between the implementer and the authorities for the licensing of spent fuel or radioactive waste management activities and facilities, although the regulator has to remain independent from the implementer. Experience has shown that licensing processes for radioactive management facilities, and in particular for geological disposal facilities, are complex and often carried out over very long time frames. They are also generally limited in number in any one country (i.e. it is unlikely that any single country will have more than a few repositories for radioactive waste or spent fuel). As such, a major challenge is for both the regulator and the implementer to build up and maintain the expertise and competencies required for licensing, constructing and operating a disposal facility. Because of this 'one off' nature, early interaction between the implementer, the nuclear regulator and perhaps other regulatory bodies can have benefits in clarifying the licensing process. An integrated approach to the regulatory process (e.g. considering all aspects at once: nuclear safety, environmental issues, health and safety, mining law, etc.) maintains clarity. In practice, regulatory interaction could be achieved by a multistep licensing process.

### 8.1.5. Sociopolitical acceptance

The attainment and maintenance of a sociopolitical consensus that enables the implementation of technical solutions for the safe management of radioactive waste and spent fuel is an open question. There is still much discussion in relation to the management of spent fuel and radioactive waste. Many States have disposal facilities for radioactive waste, but a very large number of States still need to develop them. For the management of the spent fuel there are States that have defined their policy but still have no operational disposal facility for spent fuel. The successful operational experience acquired to date becomes a fundamental element to forge and maintain society's trust in the system and thus in the operators. The first HLW disposal facility in Finland is close to operation. However, the introduction of new facilities is almost always a controversial issue both for States with this experience and for others thinking of building their first facility. This challenge is commonly tackled by means of the establishment of swift, intensive and interactive dialogue with society, with the aim of facilitating informed and responsible decision making.

Sociopolitical acceptance has been and remains today the greatest obstacle facing programmes for the disposal of spent fuel. Is it possible that the startup of the first facilities of this type over the next few years can overcome this? It would seem that this element, together with the continuous efforts to develop mechanisms for social participation in decision making, can be key to a change towards effective understanding of the need for and suitability of this type of facility.

### 8.1.6. Knowledge system sustainability

It has long been recognized that the workforce in the nuclear industry is ageing, and a considerable fraction of workers are at or near retirement age. A challenge facing the industry is how to attract enough trained and skilled staff in all disciplines, and transfer knowledge from the current generation to the new generation of workers. This issue is faced by all organizations involved, including implementers, regulators, technical support organizations, researchers, support industries, etc.

While the responsibility for the safety of radioactive waste or spent fuel management is primarily that of the waste generator, national programmes require a degree of national capability to be in place. To have the scientific, engineering and legal skills necessary to implement and regulate national programmes available requires that educational and training provisions have been made. Expertise in specialist scientific disciplines needs to be available and research capabilities are required. Bearing in mind the time frames associated with the development, operation and closure of radioactive waste/spent fuel management facilities, in particular for storage and disposal facilities, this matter of human resources is of fundamental concern to all countries, as is an understanding of the skills base and experience necessary to maintain such skills. The availability of sufficient financial resources remains a challenge for many spent fuel and radioactive waste management programmes and is particularly important for the back end activities of decommissioning and disposal. The availability and feedback of knowledge and experience on costing and financial provision for both back end activities and legacy situations is of considerable value.

Another important factor in the knowledge system is record keeping. Taking into account the long timescales of concern for the management of spent fuel and radioactive waste, special attention should be given to secure the records for the longer periods.

International cooperation, either between different organizations or under the umbrella of international organizations, has an enhanced role in sharing knowledge and in securing the safe management of spent fuel and radioactive waste.

### 8.1.7. Impact of/on newcomer programmes

The safe management of spent fuel and radioactive waste is particularly important for those States that are considering the development and implementation of a new nuclear programme or have adopted one recently. In this regard, the experience accumulated by mature nuclear programmes is offered and made available to these newcomer States by means of mechanisms to transfer knowledge overseen by international organizations. This seeks to enable States to analyse and develop the institutional and infrastructure frameworks necessary for the safe management of the responsibilities inherent to a nuclear programme, including the management of radioactive waste and spent fuel. Accordingly, most States in this situation are adopting measures for the swift enactment of the institutional, regulatory and financial framework, the training of staff, and the logistics and facilities necessary for the responsible and safe management of the back end of the nuclear cycle.

The States that have recently introduced a nuclear programme (Belarus, Türkiye, the United Arab Emirates, etc.) have been assisted by the international organizations so as to transfer the necessary experience and knowledge for the early adoption of the decisions required for the creation of a national infrastructure capable of taking on board the national and international responsibilities inherent to nuclear activity.

## 8.2. TRENDS IN RADIOACTIVE WASTE MANAGEMENT

The development of radioactive waste management programmes rests on certain global and commonly accepted principles, and on best practices that may be applied to different cases and national realities. Furthermore, we can identify approaches and developments that are being adopted and implemented in a significant number of radioactive waste management programmes to improve the safety and management of the corresponding inventories.

### 8.2.1. Radioactive waste storage and disposal needs are increasing globally

As the use of nuclear power and nuclear applications is continuing, there is a need to manage the radioactive waste produced safely. The international scientific community agrees that the definitive disposal of radioactive waste is the ultimate goal of national management programmes. Technical solutions developed or to be developed vary depending on the type, quantity and nature of the radioactive waste. Although solutions exist for the management, and in particular the disposal, of radioactive waste, the remaining topics are the duration and the requirements for the demonstration of the safety of the facilities and activities, as well as public acceptance, and this impacts on the implementation of facilities and activities — in particular disposal facilities. In recent years, there has been increasing progress in the development and improvement of legislative and regulatory frameworks, political and social dialogue, and in technical development with the aim of enabling the implementation of policies and strategies that facilitate the adoption and development of measures and actions to build facilities for final waste disposal. However, it is also often the case that the adoption of these measures requires time for their implementation and hence, in the vast majority of cases, it becomes necessary to rely on storage capabilities for radioactive waste until the final disposal route becomes available.

This trend applies to many of the national programmes that either do not have final disposal capabilities for each class of radioactive waste or currently have them but at an insufficient level to handle the entire inventory that has been/will be generated.

Because of the need to manage decommissioning of waste for its safe disposal, but also because of additional needs (e.g. the management of waste produced outside of regulated activities or post-accident), along with the improvement of the efficiency of existing capabilities, many existing facilities for the storage and disposal of radioactive waste are being adapted and improved. These improvements include the introduction of new techniques and technologies, such as facilities for the decontamination of waste, metal smelting plants and the use of new containers and configurations for the storage and disposal of this type of waste [58, 59]. This trend also impacts on the design of new facilities that adopt the lessons learned in these more mature programmes.

### 8.2.2. Management of VLLW and LLW (including disposal) as an industrial practice

Management of VLLW or LLW, which includes, inter alia, sorting, treatment and conditioning, is a question that has been resolved from a technical and practical perspective. Technologies exist for the safe management of all the streams of VLLW or LLW, with some specific exceptions, in addition to the routes and facilities for final management. Considering that the VLLW class was only introduced in the early 2000s, this concept has matured rapidly. Many States have established national systems and specific facilities to manage this type of waste and many of them have one or more disposal facilities to implement these programmes [60]. We should state that the disposal facilities in most States are centralized and unique, and fall under the responsibility of operators that, in most cases, are government agencies, although cases of private operators also exist. In some States (China, the Russian Federation, the USA), different disposal facilities also exist, which can be because of the size of the nuclear programme or because of the size of and distances within the State itself. However, it should be kept in mind that even in the case of such industrial practices, the accuracy and reliability of information has still to be addressed and demonstrated.

### 8.2.3. Disposal of ILW and HLW as a promising challenge

Although a solution for disposal of ILW and HLW has not been implemented yet, several countries are well advanced, some have been granted licences for construction and operation of facilities, and it is reasonably expected that Finland will have an operational HLW disposal facility in operation in the next few years. It has to be taken into account that these solutions require a long period of construction and operation, which should not bias the perception that those wastes are and will be safely managed. States and international organizations continue to work on this matter, which is facing severe social acceptance difficulties in most States, with achieving and maintaining a 'social licence' continuing to be one of the most challenging aspects of these processes. In this regard, efforts are geared toward the development and implementation of mechanisms for early and continuous participation in decision making, which, in the end, helps facilitate the construction and operation of this type of facility.

### 8.2.4. NORM waste as an emerging question

Different industries not related to the nuclear fuel cycle can generate NORM waste, which has been a subject of growing interest in recent years. One of the reasons for this is that the waste volumes produced by these industries can be significant. Taking into account the long time frames for operation, the accumulated inventories can be massive. In these cases, the factors for consideration include the reappraisal of their radiological, health and environmental risk, the geographical location of the industries concerned, and the changing social perception of these activities and how to manage them. NORM waste management is usually addressed with common approaches that include a combination of proven technical solutions, together with active dialogue in decision making that includes social stakeholders and interested parties in fields that are not just related to nuclear activity. This extended scope includes health and environmental authorities, and industrial corporations that were initially outside of the scope of nuclear activities.

### 8.2.5. Radioactive waste from decommissioning is the main waste stream in many States

As indicated in Section 3, concerning sources of spent fuel and radioactive waste, these are generated by a combination of activities that primarily include the operation of nuclear facilities and any other facilities that use radioactive materials for different regulated activities, such as the nuclear cycle, but also other uses in different industries, research and medicine. Over recent decades, and very significantly in the last few years, the end of operation of a large number of these facilities is generating significant inventories of different classes and types of decommissioning radioactive waste that have become the most important stream from a quantitative perspective. This situation, shared by different countries (Germany, Italy, Japan, Spain, the USA, etc.) has driven the development of specific solutions concerning such aspects as characterization, treatment and conditioning, and the associated logistics. In this regard, a general concurrence can be pointed to in identifying the questions associated with this waste, although the solutions adopted in each programme and project are more specific to their characteristics and contexts.

### 8.2.6. Waste hierarchy and circular economy becoming more important

In recent decades, the world has become more aware of the importance of issues related to environmental awareness and sustainability. In the nuclear field, since the early years, one of the key priorities for the management of radioactive waste has been minimization of the quantities and volumes to be managed. The reuse of equipment and materials and/or their recycling, as phases prior to their consideration as waste, are processes in use at all relevant facilities and in their programmes to manage radioactive materials and waste. There are increasing efforts to reduce the waste to be managed to the minimum possible, along with initiatives to recover radioactive materials that may be subject to reuse or recycling. These practices are commonly used in the operation of facilities, where radioactive material/waste is managed, as well as in planning the decommissioning projects, where the generation of candidate materials to be classified as waste can be significant in mass as in volumes.

### 8.2.7. Clearance as a key element in radioactive waste management

The option to implement clearance of radioactive waste that falls under the threshold for classification offers very interesting opportunities to reduce the volume of the inventories to be managed. To this end, a growing number of States have developed their regulatory frameworks by establishing the limits and conditions under which this practice may be implemented. These legislative and regulatory enactments are backed by the initiatives of competent international organizations in the field of radiological protection — the EC, the IAEA, and the International Commission on Radiological Protection.

These developments have a particular impact on the programmes and projects for the decommissioning of nuclear facilities as they facilitate the safe and significant reduction of the inventories of radioactive waste from decommissioning to be managed and disposed of at specific facilities.

### 8.2.8. Need for innovation: new techniques and technologies for greater efficiency

In addition to the global extension of the practice of clearance, other means are being used to achieve greater efficiency in radioactive waste management programmes that can even lead to an improvement in their alignment with the principles of the circular economy. In relation to this aspect, innovation in technologies and practices seems to offer a margin to improve the efficacy and efficiency of these activities.

This aspect of the initiatives of the operators themselves in collaboration with entities from the world of research — based on experience and the needs of final users — is supported by programmes and activities promoted by international organizations.

### 8.2.9. Specific challenges for States with small inventories

The experience of more mature programmes shows that the safe management of radioactive waste is a question that requires the convergence of a series of elements: suitable legislative and regulatory frameworks, access to knowledge and technology, availability of facilities and infrastructures, qualified personnel, financial resources and a sociopolitical consensus. These are essential elements for safe management, as set out in Section 4, the implementation and adaptation of which require an effort at a demanding and continuous State level. The attainment of these conditions in States with small inventories of radioactive waste is even more complex, since such other elements, such as prioritization of the needs of the State and funding limitations or lack of awareness of social and environmental importance, come into play. Despite these difficulties, it has been observed that a growing number of States have been completing and modernizing their structures and infrastructure, including the definition and establishment of national policies and the creation of independent regulatory bodies.

## 8.3. TRENDS IN SPENT FUEL MANAGEMENT

### 8.3.1. More Member States supporting the open cycle than the closed cycle for spent fuel management. Closed cycle supported by Member States with large programmes and foreseeable continuity

The choice between open or closed cycle for management of spent fuel is a strategic question that tends to be adopted at a State level (Section 5.1.1). In some cases, an initial decision in favour of one or the other has been revised over time, resulting in the adoption of a different option from the original one. In both cases, the final aspect of the management involves the geological disposal of those radioactive materials with no ultimate use envisaged.

Some States have not yet decided which option is the most suitable for their programme. Many States opt for an open cycle and some for a closed cycle. In this regard, we should state that most of the larger programmes (China, France, India, Japan and the Russian Federation) have opted for the closed cycle because these States are committed to the continuity of their nuclear programmes in the long term. As regards the open cycle, this is the option chosen by most of the small and medium sized nuclear programmes (Finland, Spain, Sweden and Switzerland).

### 8.3.2. Implementation of disposal programmes — news from Finland, France, Sweden, and Switzerland

Deep disposal in stable geological formations is the internationally well-established approach, supported by the scientific community, for the safe confinement of spent fuel and other long-lived radioactive waste, as its radiological characteristics can be managed by this solution. Although this solution has not yet been implemented, several countries are well advanced. Based on the national reports to the Joint Convention, it can be concluded that owing to several revisions of strategies, the programmes have suffered decades of delays in relation to the dates initially set. However, over the last two years, two elements seem to be altering this trend: the substantial progress made on several national plans, with cases where these facilities are expected to enter into service in next ten years — Finland and Sweden — and the revisions of calendars for several national programmes that have reduced their initial deadlines for introduction.

### 8.3.3. Global need to increase storage capacities

Following several decades in operation, the pool of nuclear reactors has generated a significant inventory of spent fuel, which increases yearly; accordingly, temporary storage is a need that goes beyond the choice of an open or closed cycle. Over recent years, many States have needed to extend their temporary storage capacity to accommodate these inventories by means of the extension of existing facilities or the construction of supplementary facilities.

### 8.3.4. Wet versus dry storage

Although some countries have opted to extend wet storage capacities by means of the construction of new storage pools (Finland, Sweden), it is more common for these to be dry storage facilities under the different technical existing types.

### 8.3.5. Individual storage versus consolidated/centralized storage

Furthermore, the trend is for individual rather than centralized storage facilities. Although some States have centralized storage facilities (the Kingdom of the Netherlands, Sweden, Switzerland, etc.), most States propose management strategies that involve individual temporary storage (Japan, Republic of Korea, Spain, the USA, etc.), often as a result of the difficulties in forging a social consensus for the construction of centralized storage facilities.

### 8.3.6. Interface between storage and disposal

The progress made by the most advanced national disposal plans has led to a growing awareness of and interest in the institutional and technical aspects relating to the interface between storage and disposal phases. This trend is under discussion at different international and national forums with the aim of facilitating a common understanding of the necessary developments.

### 8.3.7. Specific challenges for Member States with small inventories — including research reactors

Those States that have small inventories of spent fuel resulting from the operation of research reactors or small fleets of powerful reactors face additional challenges to the large programmes. In these cases, difficulties often exist in the allocation of technical and financial resources, while the management programme itself may be seen as a lower priority.

The repatriation of spent fuel from research reactors has been a common practice that is now declining. In the case of small fleets of powerful reactors, different approaches have been observed that include an unclear definition of the strategy (Mexico) and reprocessing (Italy), with several States promoting international or regional agreements for the disposal of spent fuel and high level waste (the Kingdom of the Netherlands, Slovenia, etc.).

In many cases, these States are backed by the action of international organizations to develop and integrate the resources and capacities needed for safe management.

### 8.3.8. Compatibility between new nuclear developments and designs (small modular reactor, Generation IV) and current management routes

Some of these new, and other already existing, national nuclear programmes (Canada, the UK, etc.) are considering the construction of nuclear reactors with novel designs. Their technical characteristics aim to employ new types of nuclear fuel that impact on the nature and properties of the spent fuel and radioactive waste that they generate. This question has generated growing interest that is being channelled by means of early dialogue with the agents involved (developers, authorities, operators), geared to taking these characteristics into account in the design of new management facilities and the modification of existing facilities.

# 9. CONCLUSIONS

This publication provides an overview of the status of spent fuel and radioactive waste management globally and presents global estimates of the amounts of residual radioactive material accumulated by nuclear activities. Significant progress has been achieved, particularly during the last 10–20 years, in the treatment, conditioning and storage of spent fuel and radioactive waste and in developing national inventories. Radioactive waste and spent fuel are safely managed all over the world, and great progress has been made on a global scale in providing more transparent and credible information.

There has also been progress in emplacing certain types of radioactive waste, as a proportion of VLLW and LLW has been disposed of using well known solutions. However, spent fuel, HLW and the major part of ILW remain in safe storage as disposal solutions are delayed. Research undertaken over several decades has progressed to the point that at least three DGRs should start operation in the next ten years. All three are located in Europe. For countries with small inventories, the development of disposal facilities can be a challenge, and not only for spent fuel or HLW waste, and for this reason there has been more discussion about the possible sharing of efforts.

The data presented in Section 6 provide a comprehensive overview and a best available estimate of the amounts of spent fuel and radioactive waste that currently exist in the world. The main source of information used for inventories is Spent Fuel and Radioactive Waste Information System (SRIS), as well as the reports submitted by the Contracting Parties to the Joint Convention.

Worldwide, there is an estimated 300 000 t HM of spent fuel in storage. Currently, approximately 7000 t HM are discharged every year. There is a decline in fuel management by reprocessing, as the UK is in the process of stopping reprocessing activities. This trend may change in the future with the development of nuclear energy and associated fuel cycle facilities in countries such as China. Therefore, special attention should be paid to setting up storage capacities and keeping them safe over time, with the storage period generally being several decades. This also applies for HLW packages from reprocessing. Some countries have decided to use centralized storage, while others store spent fuel on-site. Approximately 70% of spent fuel is currently stored in pools, but a new development is that most of the new spent fuel storage facilities are dry storages.

The overall worldwide generated volume of solid radioactive waste at the end of 2019 was approximately 35 million m$^3$. Most of it this waste (92%) is VLLW or LLW, and 30 million m$^3$ have already been disposed of, and while a further 5.6 million m$^3$ are in storage awaiting final disposal.

The available data are sufficient to provide a clear representation of the global situation in terms of the overall challenge represented by the radioactive waste that currently exists, and to provide an indication of the challenges that will arise in the future as facilities still in operation, or planned, come to the end of their useful lives. There remain some uncertainties about total global amounts of radioactive waste, as information about the spent fuel and radioactive waste inventories of all countries is not available. Additional uncertainty in the aggregated data for individual waste classes results from the need to convert data presented according to national classification systems into a common system based on the waste classification scheme in GSG-1 [4]. Finally, uncertainty also arises from the need to present waste quantities according to the anticipated 'as disposed' volumes.

Owing to the age of many nuclear power plants, decommissioning will become an increasingly important activity. The first generation of nuclear power plants are reaching the end of their design lives. This will be reinforced by changes in nuclear policy in some countries that require the shutdown of reactors. In addition, many countries are now making concerted efforts to clean up past nuclear legacy sites. Therefore, the availability of disposal routes, in particular for VLLW, will become increasingly important. In parallel, as disposal of any waste should not be the most preferred option, and as disposal facilities for radioactive waste and/or spent fuel are rare assets, there is the challenge of promoting and implementing waste minimization techniques or recycling after decontamination, in an overall optimization approach.

The availability of disposal routes is also highly important for the management of waste generated by nuclear accidents. In Japan and Ukraine, significant progress has been made towards the remediation of damaged nuclear facilities (e.g. implementation of shelters, retrieval of spent fuel, etc.) and the surrounding areas (decontamination). These works have generated large amounts of liquid and solid waste that have to be managed with a long term perspective.

Disused sealed radioactive sources (DSRSs) do not present a challenge with respect to their volumes. However, they present a safety issue owing to their quantity (some millions) and because they are widely distributed around the world for industrial uses [61]. Some countries already have a system of follow-up of sealed sources in order to enable and motivate their recovery after use.

# REFERENCES


[1]     INTERNATIONAL ATOMIC ENERGY AGENCY, Status and Trends in Spent Fuel and Radioactive Waste Management, IAEA Nuclear Energy Series No. NW-T-1.14, IAEA, Vienna (2018).

[2]     INTERNATIONAL ATOMIC ENERGY AGENCY, Status and Trends in Spent Fuel and Radioactive Waste Management, IAEA Nuclear Energy Series No. NW-T-1.14 (Rev.1), IAEA, Vienna (2022).

[3]     Joint Convention on the Safety of Spent Fuel Management and on the Safety of Radioactive Waste Management, INFCIRC/546, IAEA, Vienna (1997).

[4]     INTERNATIONAL ATOMIC ENERGY AGENCY, Classification of Radioactive Waste, IAEA Safety Standards Series No. GSG-1, IAEA, Vienna (2009).

[5]     Council Directive 2011/70/Euratom of July 2011 Establishing a Community framework for the responsible and safe management of spent fuel and radioactive waste, Official Journal of the European Union, No. L199, Publications Office of the European Union, Luxembourg (2011).

[6]     EUROPEAN COMMISION, Report from Commission to the Council and the European Parliament on progress of implementation of Council Directive 2011/70/Euratom and an inventory of radioactive waste and spent fuel present in the Community's territory and the future prospects, European Commission, Brussels (2019).

[7]     NUCLEAR ENERGY AGENCY, Nuclear Energy Data 2019, NEA No. 7474, OECD Publishing, Paris (2020), https://doi.org/10.1787/1786b86b-en-fr

[8]     Seventh Review Meeting of the Contracting Parties to the Joint Convention on the Safety of Spent Fuel Management and on the Safety of Radioactive Waste Management, Summary Report, JC/RM7/04/Rev.2, IAEA, Vienna (2022).

[9]     EUROPEAN ATOMIC ENERGY COMMUNITY, FOOD AND AGRICULTURE ORGANIZATION OF THE UNITED NATIONS, INTERNATIONAL ATOMIC ENERGY AGENCY, INTERNATIONAL LABOUR ORGANIZATION, INTERNATIONAL MARITIME ORGANIZATION, OECD NUCLEAR ENERGY AGENCY, PAN AMERICAN HEALTH ORGANIZATION, UNITED NATIONS ENVIRONMENT PROGRAMME, WORLD HEALTH ORGANIZATION, Fundamental Safety Principles, IAEA Safety Standards Series No. SF-1, IAEA, Vienna (2006), https://doi.org/10.61092/iaea.hmxn-vw0a

[10]    INTERNATIONAL ATOMIC ENERGY AGENCY, Nuclear Power Reactors in the World, Reference Data Series No. 2, IAEA, Vienna (2020).

[11]    INTERNATIONAL ATOMIC ENERGY AGENCY, Guidebook on Spent Fuel Storage Options and Systems, Technical Reports Series No. 240, IAEA, Vienna (2022).

[12]    INTERNATIONAL ATOMIC ENERGY AGENCY, Governmental, Legal and Regulatory Framework for Safety, IAEA Safety Standards Series No. GSR Part 1 (Rev. 1), IAEA, Vienna (2016).

[13]    INTERNATIONAL ATOMIC ENERGY AGENCY, Code of Conduct on the Safety and Security of Radioactive Sources, IAEA, Vienna (2004).

[14]    INTERNATIONAL ATOMIC ENERGY AGENCY, Guidance on the Import and Export of Radioactive Sources, IAEA, Vienna (2012).

[15]    INTERNATIONAL ATOMIC ENERGY AGENCY, Guidance on the Management of Disused Radioactive Sources, IAEA, Vienna (2018).

[16]    NATIONAL ACADEMIES OF SCIENCES, ENGINEERING, MEDICINE, Radioactive Sources: Applications and Alternative Technologies, National Academies Press, Washington, DC (2021).

[17]    INTERNATIONAL ATOMIC ENERGY AGENCY, Policies and Strategies for Radioactive Waste Management, IAEA Nuclear Energy Series No. NW-G-1.1, IAEA, Vienna (2009).

[18]    INTERNATIONAL ATOMIC ENERGY AGENCY, Framework and Challenges for Initiating Multinational Cooperation for the Development of a Radioactive Waste Repository, IAEA Nuclear Energy Series No. NW-T-1.5, IAEA, Vienna (2016).

[19]    NUCLEAR ENERGY AGENCY, The Evolving Role and Image of the Regulator in Radioactive Waste Management: Trends Over Two Decades, NEA No. 7083, OECD Publishing, Paris (2012).

[20]    INTERNATIONAL ATOMIC ENERGY AGENCY, Options for Management of Spent Nuclear Fuel and Radioactive Waste for Countries Developing New Nuclear Power Programmes, IAEA Nuclear Energy Series No. NW-T-1.24 (Rev. 1), IAEA, Vienna (2018).

[21]     INTERNATIONAL ATOMIC ENERGY AGENCY, IAEA Nuclear Safety and Security Glossary, IAEA, Vienna (2022),
https://doi.org/10.61092/iaea.rrxi-t56z

[22]     UNITED NATIONS INDUSTRIAL DEVELOPMENT ORGANIZATION (UNIDO), "Global consultations on circular economy 2022, Background note" (2022),
https://www.unido.org/sites/default/files/files/2022-09/Background%20note%20for%20the%20global%20consultations%20on%20circular%20economy%202022.pdf

[23]     INTERNATIONAL ATOMIC ENERGY AGENCY, Stakeholder Involvement Throughout the Life Cycle of Nuclear Facilities, IAEA Nuclear Energy Series No. NG-T-1.4, IAEA, Vienna (2011).

[24]     INTERNATIONAL NUCLEAR SAFETY GROUP, Stakeholder Involvement in Nuclear Issues, INSAG Series No. 20, IAEA, Vienna (2006).

[25]     INTERNATIONAL ATOMIC ENERGY AGENCY, Communications on Nuclear, Radiation, Transport and Waste Safety: A Practical Handbook, IAEA-TECDOC-1076, IAEA, Vienna (1999).

[26]     INTERNATIONAL ATOMIC ENERGY AGENCY, Factors Affecting Public and Political Acceptance for the Implementation of Geological Disposal, IAEA-TECDOC-1566, IAEA, Vienna (2007).

[27]     UNITED NATIONS ECONOMIC COMMISSION FOR EUROPE, Convention on Access to Information, Public Participation in Decision-Making and Access to Justice in Environmental Matters, UNECE, Geneva (1998).

[28]     UNITED NATIONS ECONOMIC COMMISSION FOR EUROPE, Convention on Environmental Impact Assessment in a Transboundary Context, UNECE, Geneva (2017).

[29]     INTERNATIONAL ATOMIC ENERGY AGENCY, Disposal of Radioactive Waste, IAEA Safety Standards Series No. SSR-5, IAEA, Vienna (2011).

[30]     INTERNATIONAL ATOMIC ENERGY AGENCY, Design Principles and Approaches for Radioactive Waste Repositories, IAEA Nuclear Energy Series No. NW-T-1.27, IAEA, Vienna (2020).

[31]     INTERNATIONAL ATOMIC ENERGY AGENCY, Pressurized Heavy Water Reactor Fuel: Integrity, Performance and Advance Concepts, IAEA-TECDOC (CD-ROM) No. 1751, IAEA, Vienna (2014).

[32]     INTERNATIONAL ATOMIC ENERGY AGENCY, Status and Advances in MOX Fuel Technology, Technical Reports Series No. 415, IAEA, Vienna (2003).

[33]     UNITED STATES DEPARTMENT OF ENERGY, A Historical Review of the Safe Transport of Spent Nuclear Fuel, USDOE, Washington, DC (2016).

[34]     INTERNATIONAL ATOMIC ENERGY AGENCY, Operation and Maintenance of Spent Fuel Storage and Transportation Casks/Containers, IAEA-TECDOC-1532, IAEA, Vienna (2007).

[35]     INTERNATIONAL ATOMIC ENERGY AGENCY, Regulations for the Safe Transport of Radioactive Material, IAEA Safety Standards Series No. SSR-6 (Rev.1), IAEA, Vienna (2018),
https://doi.org/10.61092/iaea.ur52-my9o

[36]     NUCLEAR ENERGY AGENCY, The Safety of Long-Term Interim Strorage Facilities in NEA Member Countries, OECD Publishing, Paris (2017).

[37]     INTERNATIONAL ATOMIC ENERGY AGENCY, Scientific and Technical Basis for the Geological Disposal of Radioactive Wastes, Technical Reports Series No. 413, IAEA, Vienna (2003).

[38]     NATIONAL RESEARCH COUNCIL, Disposition of High-Level Waste and Spent Nuclear Fuel: The Continuing Societal and Technical Challenges, National Academies Press, Washington, DC (2001).

[39]     NUCLEAR ENERGY AGENCY, International Conference on Geological Repositories 2016, NEA No. 7345, OECD Publishing, Paris (2017).

[40]     INTERNATIONAL ATOMIC ENERGY AGENCY, The Use of Scientific and Technical Results from Underground Research Laboratory Investigations for the Geological Disposal of Radioactive Waste, IAEA-TECDOC-1243, IAEA, Vienna (2001).

[41]     NUCLEAR ENERGY AGENCY, Underground Research Laboratories (URL), NEA No. 78122, OECD Publishing, Paris (2013).

[42]     INTERNATIONAL ATOMIC ENERGY AGENCY, Available Reprocessing and Recycling Services for Research Reactor Spent Nuclear Fuel, IAEA Nuclear Energy Series No. NW-T-1.11, IAEA, Vienna (2017).

[43]  EUROPEAN COMMISSION, FOOD AND AGRICULTURE ORGANIZATION OF THE UNITED NATIONS, INTERNATIONAL ATOMIC ENERGY AGENCY, INTERNATIONAL LABOUR ORGANIZATION, OECD NUCLEAR ENERGY AGENCY, PAN AMERICAN HEALTH ORGANIZATION, UNITED NATIONS ENVIRONMENT PROGRAMME, WORLD HEALTH ORGANIZATION, Radiation Protection and Safety of Radiation Sources: International Basic Safety Standards, IAEA Safety Standards Series No. GSR Part 3, IAEA, Vienna (2014),
https://doi.org/10.61092/iaea.u2pu-60vm

[44]  Council Directive 2013/59/Euratom, of 5 December 2013 laying down basic safety standards for protection against the dangers arising from exposure to ionising radiation, and repealing Directives 89/618/Euratom, 90/641/Euratom, 96/29/Euratom, 2013, EC, Luxembourg (2013).

[45]  Council Regulation (Euratom) No. 1493/93 of 8 June 1993, on shipments of radioactive substances between Member States, Official Journal of the European Communities, no. L148, Office for Official Publications of the European Communities, Luxembourg (1993).

[46]  INTERNATIONAL ATOMIC ENERGY AGENCY, Predisposal Management of Radioactive Waste from the Use of Radioactive Material in Medicine, Industry, Agriculture, Research and Education, IAEA Safety Standards Series No. SSG-45, IAEA, Vienna (2019).

[47]  ENSREG WG2, Guidelines for Member States reporting on Article 14.1 of Council Directive 2011/70/Euratom, EC, Luxembourg (2018).

[48]  NATIONAL ACADEMIES OF SCIENCES, ENGINEERING, MEDICINE, Reducing the Use of Highly Enriched Uranium in Civilian, National Academies Press, Washington, DC (2016).

[49]  NUCLEAR ENERGY AGENCY, The Economics of the Back End of the Nuclear Fuel Cycle, NEA No. 7061, OECD Publishing, Paris (2013).

[50]  Management of Spent Fuel from Nuclear Power Reactors, (Proc. Series International Atomic Energy Agency), IAEA, Vienna (2020).

[51]  INTERNATIONAL ATOMIC ENERGY AGENCY, Spent Fuel Performance Assessment and Research, IAEA-TECDOC-1680, IAEA, Vienna (2012).

[52]  INTERNATIONAL ATOMIC ENERGY AGENCY, Spent Fuel Performance Assessment and Research, IAEA-TECDOC-1771, IAEA, Vienna (2015).

[53]  INTERNATIONAL ATOMIC ENERGY AGENCY, Behaviour of Spent Power Reactor Fuel During Storage, IAEA-TECDOC-1862, IAEA, Vienna (2019).

[54]  NUCLEAR ENERGY AGENCY, "Building Constructive Dialogues Between Regulators and Implementers During the Pre-Licensing Phase of Deep Geological Repository Development", Radioactive Waste Management and Decommissioning NEA/RWM/R(2022)4, Paris, January 2024.

[55]  INTERNATIONAL ATOMIC ENERGY AGENCY, Predisposal Management of Radioactive Waste from Nuclear Power Plants and Research Reactors, IAEA Safety Standards Series No. SSG-40, IAEA, Vienna (2016).

[56]  INTERNATIONAL ATOMIC ENERGY AGENCY, Predisposal Management of Radioactive Waste from Nuclear Fuel Cycle Facilities, IAEA Safety Standards Series No. SSG-41, IAEA, Vienna (2016).

[57]  INTERNATIONAL ATOMIC ENERGY AGENCY, Milestones in the Development of a National Infrastructure for Nuclear Power, IAEA Nuclear Energy Series No. NG-G-3.1 (Rev. 1), IAEA, Vienna (2015),
https://doi.org/10.61092/iaea.hff3-zuam

[58]  INTERNATIONAL ATOMIC ENERGY AGENCY, Application of Thermal Technologies for Processing of Radioactive Waste, IAEA-TECDOC-1527, IAEA, Vienna (2006).

[59]  INTERNATIONAL ATOMIC ENERGY AGENCY, Selection of Technical Solutions for the Management of Radioactive Waste, IAEA-TECDOC-1817, IAEA, Vienna (2017).

[60]  NATIONAL ACADEMIES OF SCIENCES, ENGINEERING, AND MEDICINE, Low-Level Radioactive Waste Management and Disposition: Proceedings of a Workshop, National Academies Press, Washington, DC (2017).

[61]  INTERNATIONAL ATOMIC ENERGY AGENCY, Management of Disused Sealed Radioactive Sources, IAEA Nuclear Energy Series No. NW-T-1.3, IAEA, Vienna (2014).

# SUMMARY OF WASTE CLASSIFICATION SYSTEMS

| COUNTRY | CLASSIFICATION SYSTEM |
|---|---|

**Argentina [I–1]**

Uses the IAEA classification system

**Australia [I–2]**

Uses the IAEA classification system

**Austria [I–3]**

*Transition radioactive waste* — type of radioactive waste (mainly from medical origin) that will decay within the period of temporary storage and may then be suitable for management outside of the regulatory control system subject to compliance with clearance levels. (Waste in the transition phase, e.g. short lived decay waste from medical applications containing I-125, is put into decay storage prior to free release.)

*Low and intermediate level waste (LILW)* — in this, the concentration of radionuclides is such that generation of thermal power during its disposal is sufficiently low. These acceptable thermal power values are site specific following safety assessments.

— *Short lived waste (LILW-SL):* this category includes radioactive waste with nuclide half-lives less than or equal to those of Cs-137 and Sr-90 (~30 years) with a restricted alpha long lived radionuclide concentration (limitation of long lived alpha emitting radio nuclides to 4000 Bq/g in individual waste packages and to an overall average of 400 Bq/g in the total waste volume).
— *Long lived waste (LILW-LL):* long lived radionuclides and alpha emitters whose concentration exceeds the limits for short lived waste.

*High level waste (HLW)* — does not arise in Austria

**Belarus [I–4]**

By specific activity, liquid and solid radioactive wastes are divided into four categories: very low level, low level, intermediate level and high level. Liquid radioactive wastes are divided into three categories: low level, intermediate level and high level.

*Solid:*
— Very low level: $<10^3$ β, $<10^2$ α, $<10^1$ transuranic (Bq/g)
— Low level: $10^3$–$10^4$ β, $10^2$–$10^3$ α, $10^1$–$10^2$ transuranic (Bq/g)
— Intermediate level: $10^4$–$10^7$ β, $10^3$–$10^6$ α, $10^2$–$10^5$ transuranic (Bq/g)
— High level: $>10^7$ β, $>10^6$ α, $>10^5$ transuranic (Bq/g)
*Liquid:*
— Low level: $<10^3$ β, $<10^2$ α, $<10^1$ transuranic (Bq/g)
— Intermediate level: $10^3$–$10^7$ β, $10^2$–$10^6$ α, $10^1$–$10^5$ transuranic (Bq/g)
— High level: $>10^7$ β, $>10^6$ α, $>10^5$ transuranic (Bq/g)

**Belgium**
[I–5]

*Category A* — waste is short lived, low level and intermediate level conditioned waste containing limited quantities of long lived radionuclides. It poses a risk to people and the environment for several hundreds of years. It can be considered for surface or near surface disposal. It corresponds to low level waste in the IAEA 2009 classification. The radiological criteria and limits for the category A waste will be defined in the safety report and licensing conditions for the planned disposal facility in Dessel.

*Category B* — waste is low level and intermediate level conditioned waste contaminated with such quantities of long lived radionuclides that it poses a risk to people and the environment for several tens to several hundreds of thousands of years in some cases. Note that sealed sources that have to be managed as radioactive waste end up in category B after treatment and conditioning. Its thermal power is potentially significant at the time of its conditioning, but it will emit too little heat after the storage period to be classified as category C waste. It corresponds to intermediate level waste in the IAEA 2009 classification.

*Category C* — waste is high level conditioned waste containing large quantities of long lived radionuclides that, like category B waste, poses a risk for several tens to several hundreds of thousands of years in some cases. After the period currently considered for its storage (~60 years of cooling required, in the event of subsequent disposal in poorly indurated clay), its thermal power still causes a significant increase in the temperature of the repository's host rock. It corresponds to high level waste in the IAEA 2009 classification. Category C waste includes vitrified waste from the reprocessing of spent fuel from commercial nuclear power plants (and from the BR2 research reactor) and non-reprocessed spent fuel declared as waste, except for certain fuels from research reactors, which belong to category B.

|                   | Low activity | Medium activity | High activity |
| ----------------- | :----------: | :-------------: | :-----------: |
| Short lived waste | A            | A               | C             |
| Long lived waste  | B            | B               | C             |

The above waste categories are further subdivided in waste classes and waste streams.

**Bosnia and Herzegovina**
[I–6]

Uses the IAEA classification system

**Botswana**
[I–7]

*Cleared material/waste* — waste containing levels of radionuclides at concentrations less than the clearance levels established by the Radiation Protection Inspectorate.

*Low level (short lived)/decay waste* — low level radioactive waste containing short lived radionuclides only (e.g. with half-lives <10 days) that will decay to clearance levels within three years after the time of its generation.

*Low and intermediate level, short lived waste (LILW-SL)* — waste that will not decay to clearance levels within three years and contains beta/gamma emitting radionuclides with half-lives <30 years and/or alpha emitting radionuclides with an activity <400 Bq/g and a total activity <4000 Bq in each waste package.

*Low and intermediate level, long lived waste (LILW-LL)* — radioactive waste containing radionuclides with concentrations above those for LILW-SL, but which does not generate heat at >2 kW/m³ of waste.

CLASSIFICATION SYSTEM

---

Brazil
[I–8]

Uses the IAEA classification system

Bulgaria
[I–9]

*Category 1* — transitional waste that contains small concentrations of safety significant radionuclides so that it does not require provisions for radiation protection or does not need a high level of containment and isolation; this category of waste is additionally subdivided into:

> *Category 1a* — exempt waste

*Category 2* — low and intermediate level waste (LILW). Because of its radionuclide content, LILW requires robust isolation and containment but no special measures for heat removal during its storage and disposal; this category of waste is additionally subdivided into:

> *Category 2a* — low and intermediate level waste containing mainly short lived radionuclides (with a half-life no longer than the half-life of Cs-137) as well as long lived radionuclides at significantly lower levels of activity, limited for the long lived alpha emitters to $<4 \times 10^6$ Bq/kg for each individual package and a maximum average for all packages in the respective facility of $4 \times 10^5$ Bq/kg; for such waste, reliable isolation and containment is required for a period of up to several hundred years.

> *Category 2b* — low and intermediate level waste containing long lived radionuclides at activity levels of long lived alpha emitters, exceeding the limits for category 2a.

*Category 3* — high level waste: with concentration of radionuclides at which heat removal has to be considered in its storage and disposal; a higher level of isolation and containment compared to low and intermediate level waste is needed through disposal in deep, stable geological formations.

Canada
[I–10]

*Uranium mine and mill waste* — waste rock and mill tailings are a specific type of radioactive waste generated during the mining and milling of uranium ore and the production of uranium concentrate. In addition to tailings, mining activities typically produce large quantities of mineralized and clean waste rock excavated to access the ore body. The tailings and mineralized waste rock contain significant concentrations of long lived radioactive elements, namely Th-230 and Ra-226.

*Low level radioactive waste (LLW)* — contains material with radionuclide content above established clearance levels and exemption quantities, and generally limited amounts of long lived activity. LLW requires isolation and containment for up to a few hundred years. LLW generally does not require significant shielding during handling and interim storage. LLW has two subcategories:

---

*Very short lived low level radioactive waste (VSLLW)* — can be stored for decay for up to a few years and subsequently cleared for release. This classification includes radioactive waste containing only short half-life radionuclides of the kind typically used for research and biomedical purposes. Examples of VSLLW are Ir-192 and Tc-99m sources, as well as industrial and medical radioactive waste that contains similar short half-life radionuclides. Generally, the main criterion for VSLLW is the half-life of the predominant radionuclides. In practice, the management of VSLLW should only be applied to radionuclides with a half-life of 100 days or less.

*Very low level radioactive waste (VLLW)* — has a low hazard potential but is nevertheless above the criteria for exemption. Long term waste management facilities for VLLW do not usually need a high degree of containment or isolation. A near surface repository with limited regulatory control is generally suitable. Typically, VLLW includes bulk material such as low activity soil and rubble, decommissioning wastes and some uranium contaminated wastes.

*Intermediate level radioactive waste (ILW)* — waste that typically exhibits sufficient levels of penetrating radiation to warrant shielding during handling and interim storage. This type of radioactive waste generally requires little or no provision for heat dissipation during its handling, transportation and long term management. However, some ILW may have heat generation implications in the short term (e.g. refurbishment waste) because of its total radioactivity level.

*High level radioactive waste (HLW)* —used (irradiated) nuclear fuel that has been declared radioactive waste or waste that generates significant heat (typically >2 kW/m$^3$) via radioactive decay. In Canada, 'irradiated nuclear fuel' or 'used nuclear fuel' are more accurate terms for spent fuel because discharged fuel is considered a waste material even when it is not fully spent. Despite the name difference, in this report the term 'spent fuel' is used to be consistent with the terminology found in the Joint Convention. Spent fuel is associated with penetrating radiation, which requires shielding. Furthermore, spent fuel contains significant quantities of long lived radionuclides, meaning that long term isolation is also required. Waste forms derived from spent fuel (e.g. nuclear fuel reprocessing wastes) can exhibit similar characteristics and may be considered HLW. Placement in deep, stable geological formations is considered the preferred option for the long term management of HLW.

| | |
|---|---|
| Chile<br>[I–11] | *Category 1* — alpha emitting radioisotopes, whatever their activity<br>*Category 2* — radioisotopes with beta and gamma emitters whose half-life is >100 days<br>*Category 3* — Radioisotopes with beta and gamma emitters whose half-life is <100 days |
| China<br>[I–12] | Uses the IAEA classification system |

---

**Croatia**
**[I–13]**

*Exempt and cleared radioactive waste* — radioactive waste that fulfils the requirements for release from regulatory control.

*Very short lived radioactive waste* — radioactive waste containing radionuclides with half-life <100 days.

*Low level radioactive waste* — radioactive waste containing radionuclides with a half-life <30 years with a restricted alpha long lived radionuclide concentration (limitation of long lived alpha emitting radionuclides to 4000 Bq/g in individual waste packages and to an overall average of 400 Bq/g in the total waste volume). The generation of thermal power in this waste is <2 kW/m$^3$.

*Intermediate level radioactive waste* — radioactive waste with activity concentrations exceeding the limits for low level radioactive waste.

*High level radioactive waste* — radioactive waste thermal power >2 kW/m$^3$.

**Cuba**
**[I–14]**

*Low and intermediate level and very short lived waste* — radioactive waste containing radionuclides with a short half-life (<100 days), which, after a period of interim storage, will decay to clearance levels.

*Low level and intermediate level and short lived waste* — radioactive waste containing radionuclides with activity levels above the clearance levels established by the regulatory body, with a half-life >100 days and <30 years, which does not generate residual heat >2 kW/m$^3$.

*Low and intermediate level and long lived waste* — radioactive waste containing radionuclides with activity levels above the clearance levels established by the regulatory body, with a half-life >30 years, which does not generate residual heat >2 kW/m$^3$.

**Cyprus**
**[I–15]**

Uses the IAEA classification system

**Czech Republic**
**[I–16]**

*Temporary radioactive waste* —after storage for at most five years shows radioactivity lower than clearance levels.

*Very low level waste* — with radioactivity higher than that of temporary radioactive waste, but does not require any special measures during disposal.

*Low level waste* — with radioactivity higher than that of temporary radioactive waste, but at the same time contains limited amounts of long lived radionuclides.

*Intermediate level waste* — contains a significant amount of long lived radionuclides, and therefore requires a higher degree of isolation from the surrounding environment than low level waste.

*High level waste* — during storage and disposal, it is necessary to take into account heat generated by decay of the contained radionuclides; the waste is processed and treated to meet the acceptance criteria and it has to be disposed in deep geological repositories several hundred metres underground.

---

Estonia
[I–17]

*Exempt waste* — waste arising from radiation practices, for which the activity and activity concentration or activity concentration on the surface is lower that the clearance levels established.

*NORM waste* — radioactive waste produced as a result of handling raw materials containing substances that contain natural radionuclides (Th-232 and U-238 and radionuclides belonging to their decay series), the specific activity of which is greater than the clearance levels established.

*Short lived waste* — radioactive waste containing radionuclides with a half-life <100 days that will decay below the clearance levels established within up to five years.

*Low and intermediate activity, short lived waste* — radioactive waste that contains beta and gamma sources with a half-life <30 years and a limited amount of long lived alpha sources (no more than 4000 Bq/g for one waste package and no more than 400 Bq/g averaged over the total waste package amounts).

*Low and intermediate activity, long lived waste* — radioactive waste containing radionuclides with a half-life >30 years with an activity concentration higher than that for low and intermediate activity short lived waste that will generate <2 kW/m$^3$ heat energy by radioactive decay.

*High level waste* — radioactive waste that generates >2 kW/m$^3$ heat energy by radioactive decay.

Finland
[I–18]

The classification system for the purpose of predisposal management of LILW from nuclear facilities, including nuclear power plants (NPPs), is based on activity concentrations. Solid and liquid waste arising from the controlled area of an NPP contain almost exclusively short lived beta and gamma emitters and are grouped into the following activity categories:

*Very low level waste (VLLW)* — refers to waste whose average activity concentration of significant radionuclides does not exceed the value of 100 kBq/kg and whose total activity does not exceed 1 TBq, α-activity <10 GBq.

*Low level waste (LLW)* — contains so little radioactivity that it can be treated without any special radiation protection arrangements. The activity concentration in the waste cannot be >1 MBq/kg.

*Intermediate level waste (ILW)* — contains radioactivity to the extent that effective radiation protection arrangements are needed when the waste is processed. As a rule, the activity concentration in the waste is from 1 MBq/kg to 10 GBq/kg.

*HLW level waste (HLW)* — spent fuel is regarded as HLW.

The classification for disposal distinguishes short lived and long lived waste accordingly:

*Short lived waste* — refers to nuclear waste of which the activity concentration after 500 years will be <100 MBq/kg in each disposed waste package and below an average value of 10 MBq/kg of waste in one emplacement room.

*Long lived waste* — refers to nuclear waste of which the activity concentration after 500 years will be >100 MBq/kg in a disposed waste package, or above an average value of 10 MBq/kg of waste in one emplacement room.

France
[I–19]

*Very short lived* — waste mainly comes from medical and non-nuclear power generating research.

*Very low level waste* — mainly comes from the decommissioning of NPPs, fuel cycle facilities and research centres and, to a lesser extent, from the operation and maintenance of these types of nuclear installation. The activity level of this waste is generally <100 Bq/g.

*Low level and intermediate level, short lived waste* — waste results mainly from the operation, maintenance and decommissioning of NPPs, fuel cycle facilities, research centres and, to a far lesser extent, from medical research activities. The level of this waste is between a few hundred and a million Bq/g.

*Low level, long lived waste* — mainly graphite waste and radium bearing waste. Graphite waste comes primarily from the old gas cooled reactor technology. Graphite waste primarily contains long lived beta radionuclides such as C-14 and Cl-36. Their activity level is ~10 000–100 000 Bq/g. Radium bearing waste, mostly from non-nuclear power generating activities, mainly contains long lived alpha emitting radionuclides and their activity level is between a few tens to a few thousand Bq/g.

*Intermediate level, long lived waste* — mainly comes from spent fuels after reprocessing and activities involved in the operation and maintenance of fuel reprocessing plants. The activity level of this waste is ~1 million to 1 billion Bq/g. The heat given off is slight or negligible.

*High level waste* — waste consists mainly of vitrified waste packages from spent fuels after reprocessing. The activity level of vitrified waste lies in the order of several billion Bq/g and it generates significant heat.

*Spent fuel* — is not generally considered to be a waste and is normally reprocessed and recycled. However, some limited amounts of spent fuel that are not compatible with the reprocessing facilities may be disposed of directly as waste.

Georgia
[I–20]

Uses the IAEA classification system

Germany
[I–21]

*Negligible heat generating radioactive waste* — radioactive waste with an average heat output of <~200 W/m$^3$ of waste (corresponding to a 3 K increase in temperature at the wall of the disposal chamber of the Konrad repository caused by decay heat from the radionuclides contained in the waste packages).

*Heat generating radioactive waste* — characterized by high activity concentrations and therefore by high decay heat output. This category includes reprocessing residues and spent fuel.

Ghana
[I–22]

Uses the IAEA classification system

Greece
[I–23]

Uses the IAEA classification system

**Hungary**
**[I–24]**

Radioactive waste producing a significant amount of heat is classified as high level waste. Heat generation is significant if it needs to be considered during the design of storage and disposal and during operation.

The rest of the radioactive wastes are low and intermediate level wastes, which are to be classified considering the following aspects:
— *Low and intermediate level radioactive waste,* in which the half-life of the radionuclides is 30 years or less, and which contains long lived alpha emitter radionuclides only in limited concentration, is considered short lived.
— *Low and intermediate level radioactive waste*, in which the half-life of the radionuclides and/or the concentration of the alpha emitter radionuclides exceed the limits of short lived radioactive waste, is considered long lived.

The classification of the radioactive waste into low and intermediate level classes will be performed based on the activity concentration and exemption activity concentration of the given radioisotope.

**Indonesia**
**[I–25]**

Solid
*Low level waste:*
— Very short half-life
— Very low level
— Relative low level

*Intermediate level waste:*
— DSRSs with a half-life <15 years
— DSRSs with a half-life of 15–30 years
— DSRS with a half-life >30 years
— Other than DSRSs

*High level waste:*
— Spent nuclear fuel

*Liquid:*
— Low active — activities of 37 kBq/m$^3$ to 0.37 GBq/m$^3$
— Intermediate active — activities of 0.37 GBq/m$^3$ to 3700 GBq/m$^3$

| | |
|---|---|
| International Waste Classification System [I–26] | *Exempt waste (EW)* — activity levels at or below clearance levels, which are based on an annual dose to members of the public of <0.01 mSv.<br><br>*Very short lived waste (VSLW)* — waste that can be stored for decay over a limited period of up to a few years and subsequently cleared from regulatory control according to arrangements approved by the regulatory body, for uncontrolled disposal, use or discharge. This class includes waste containing primarily radionuclides with very short half-lives, often used for research and medical purposes.<br><br>*Very low level waste (VLLW)* — radioactive waste considered suitable by the regulatory body for authorized disposal, subject to specified conditions, with ordinary waste in facilities not specifically designed for radioactive waste disposal.<br><br>*Low and intermediate level waste (LILW)* — activity levels above EW and thermal power $<\sim 2$ kW/m$^3$.<br><br>LILW is subdivided into:<br>— Short lived low and intermediate level waste (LILW-SL), short lived waste with long lived alpha nuclides of <400 Bq/g average or 4000 Bq/g for individual packages.<br>— Long lived low and intermediate level waste (LILW-LL), long lived concentrations above LILW-SL.<br><br>*High level waste (HLW)* — thermal power >2 kW/m$^3$ and long lived radionuclide concentrations above LILW-SL. This category includes spent fuel where it has been declared a waste. |
| Italy [I–27] | *Very short lived waste* — radioactive waste containing radionuclides with a very short half-life, of <100 days, requiring up to five years to reach activity concentrations lower than the values specified in legislation.<br><br>*Very low level waste* — radioactive waste with an activity concentration that does not meet the criteria set out for exempt waste, but <100 Bq/g, with a maximum alpha contribution of 10 Bq/g for alpha emitting long lived radionuclides.<br><br>*Low level waste* — radioactive waste that does not meet the criteria established for exempt waste and that requires containment and isolation periods of up to a few hundred years to be disposed of. This category includes radioactive waste characterized by levels of activity concentration of up to 5 MBq/g for short lived radionuclides, of up to 40 kBq/g for the long lived isotopes of nickel and of up to 400 Bq/g for long lived radionuclides.<br><br>*Intermediate level waste* — Radioactive waste with activity concentrations exceeding the values set out for low level waste, though not requiring provisions for heat dissipation during storage and disposal. This category includes waste containing long lived radionuclides that mostly requires a degree of isolation higher than that provided by near surface disposal facilities with engineered barriers, therefore requiring disposal in geological formations. This category also includes waste characterized by levels of activity concentrations of up to 400 Bq/g for alpha emitting radionuclides and mainly containing radionuclides beta/gamma emitters (even long lived), with such activity concentrations that they can be disposed of in near surface facilities with engineered barriers, provided that the level of activity concentration complies with the objectives of radiation protection established for the abovementioned surface disposal facility, such as, for instance, the waste containing activation products arising from the decommissioning of some parts of the nuclear facilities.<br><br>*High level waste* — radioactive waste with high activity concentrations, such as to generate a significant amount of heat or with high concentrations of long lived radionuclides, or both of these characteristics, which require a degree of isolation and containment for a time period of thousands of years and more. This waste requires disposal in geological formations. |

**Japan**
[I–28]

Two basic solid waste categories:

*Category 1* — final disposal is burial of radioactive waste that exceeds the criteria defined by legislation as it poses potentially significant risks to human health.

*Category 2* — all other radioactive waste, subdivided into several categories based on their disposal paths:
— *Intermediate depth disposal* is burial of radioactive waste at a depth of 70 m and up and not exceeding criteria defined by legislation.
— *Pit disposal* is burial of radioactive waste above ground or less than 70 m underground, and not exceeding criteria defined by legislation.
— *Trench disposal* is burial of radioactive waste above ground or less than 70 m underground, and not exceeding criteria defined by legislation.

**Jordan**
[I–29]

Uses the IAEA classification system

**Kazakhstan**
[I–30]

Radioactive waste is classified by activity level as follows:

*Low level waste* — waste with specific activity of <1000 Bq/kg for beta emitting nuclides, <100 Bq/kg for alpha emitting nuclides (excluding transuranium) and <10 Bq/kg for transuranium nuclides

*Intermediate waste* — waste with specific activity from 1000 Bq/kg–10 000 000 Bq/kg for beta emitting nuclides, from 100 Bq/kg–1 000 000 Bq/kg for alpha emitting nuclides (excluding transuranium) and from 10 Bq/kg–100 000 Bq/kg for transuranium nuclides.

*High level waste* — waste with specific activity of >10 000 000 Bq/kg for beta emitting nuclides, >1 000 000 Bq/kg for alpha emitting nuclides (excluding transuranium) and >100 000 Bq/kg for transuranium nuclides.

**Kingdom of the Netherlands**
[I–31]

*Exempt waste*

*Short lived waste*

*Low level and intermediate level radioactive waste LILW* (including NORM waste):
— *Category A:* alpha bearing waste
— *Category B:* beta/gamma contaminated waste from nuclear power plants
— *Category C:* beta/gamma contaminated waste from generators other than nuclear power plants with a half-life >15 years
— *Category D:* beta/gamma contaminated waste from generators other than nuclear power plants with a half-life <15 years

*High level radioactive waste (HLW)* (non-heat generating and heat generating):
— Heat generating, consists of the vitrified waste from reprocessing of spent fuel from the two nuclear power reactors in the Kingdom of the Netherlands (Borssele and Dodewaard), the spent fuel of the two research reactors (Petten and Delft) and the spent uranium targets from molybdenum production.

Non heat generating is mainly formed by reprocessing waste other than the vitrified residues. It also includes waste from research on reactor fuel and some decommissioning waste.

**Latvia**
[I–32]

Uses the IAEA classification system

Lithuania
[I–33]

*Very low level waste (VLLW)* — radioactive waste with radiological characteristic values exceeding clearance levels, but lower than the characteristics for low level waste. VLLW will be disposed in licensed landfills.

*Low and intermediate level waste (LILW)* — radioactive waste with radiological characteristics between those of very low level waste and high level waste. These may be long lived waste (LILW-LL) or short lived waste (LILW-SL). LILW-SL will be disposed of in a near surface repository (NSR) at the site .

*High level waste (HLW)* — spent fuel that will be placed in a deep geological repository owing to its significant emitting capacity of heat generated during radioactive decay or owing to the amount of long lived radionuclides contained.

Madagascar
[I–34]

*Exempt waste* — wastes whose very low level of activity allows their elimination by conventional techniques, without any particular consideration for their radioactive properties.

*Low and medium level waste* — short lived low level and intermediate level waste has to be divided into three categories according to radioactive period:
— Type 1 waste, which contains only radionuclides with a period of <6 days
— Type 2 waste, which contains radionuclides with a period of 6–71 days
— Type 3 waste, which contains radionuclides with a period of >71 days

*Long lived waste* — waste whose average activity in long lived alpha emitters exceeds that set for short lived waste.

*High activity waste* — waste with a thermal capacity of >2 kW m$^3$ and a concentration of long life emitters exceeding that set for short lived waste.

Malawi
[I–35]

Uses the IAEA classification system

Mauritius
[I–36]

Uses the IAEA classification system

Mexico
[I–37]

*Low level waste:*
— *Class A:* Radioactive waste that within a period of 100 years constitutes a level of risk that is acceptable for the population and environment
— *Class B:* Radioactive waste that within a period of 300 years constitutes a level of risk that is acceptable for the population and environment
— *Class C:* Radioactive waste that within a period of 500 years constitutes a level of risk that is acceptable for the population and environment

*Intermediate level waste* — radioactive waste whose risk remains above acceptable levels for more than 500 years

*High level waste* — radioactive wastes arising from reprocessing of spent nuclear fuel and spent nuclear fuel itself, once the Government declares it as waste

*Mixed waste* — radioactive waste mixed with other hazardous materials

*Uranium and thorium tailings* — radioactive waste arising from ore processing in a mill to extract metal content

Montenegro
[I–38]

Uses the IAEA classification system

Morocco
[I–39]

Uses the IAEA classification system

Nigeria
[I–40]

*Category I* — low level radioactive waste (<10 MBq) containing short lived radionuclides only (<50 days) that will decay to clearance levels within one year after the time of its generation.

*Category II* — low and intermediate level radioactive waste, containing radionuclides with a half-life <30 years and restricted long lived radionuclide concentrations and that is not expected to decay to clearance levels within one year from the time of its generation (limitation of longer lived alpha emitting radionuclides to 400 Bq/g individual waste packages and to an overall average of 400 Bq/g per waste packages).

*Category III* — low and intermediate level radioactive waste, containing radionuclides with a half-life >30 years and a concentration of alpha emitters exceeding the limitations for category II. This waste needs to be disposed of in deep geological facilities only.

*Category IV* — high level radioactive waste with thermal power >2 kW/m$^3$ and a concentration of alpha emitters exceeding the limitations for category III (e.g. spent fuel from research reactors). This waste needs to be disposed of in deep geological facilities only.

North Macedonia
[I–41]

*Cleared waste*

*Radioactive waste with radionuclides with a very short half-life* — radioactive waste that may be stored over a limited period of time of a few years and subsequently be released from control; this class includes radionuclides used for research and medical purposes.

*Very low activity radioactive waste*

*Low activity radioactive waste* — radioactive waste that is above release levels, but with limited amounts of long lived radionuclides; such radioactive waste requires robust isolation and containment for periods of up to a few hundred years and is suitable for disposal in engineered near surface facilities; low activity radioactive waste may include short lived radionuclides at higher levels of activity concentration and long lived radionuclides, but only at relatively low levels of activity concentration.

*Intermediate activity radioactive waste* — radioactive waste that, because of its content, particularly long lived radionuclides, requires a greater degree of containment and isolation than that provided by near surface disposal; such waste needs only limited provision for heat dissipation during its storage and disposal; it may contain long lived radionuclides, in particular alpha emitters, that will not decay to a level of activity concentration acceptable for near surface disposal; such radioactive waste requires disposal at greater depths, in the order of tens of metres to a few hundred metres.

*High activity radioactive waste* — such radioactive waste requires disposal in stable geological formations several hundred metres below the surface.

Paraguay
[I–42]

*Very short half-life waste* — waste that has to be stored for decay for a limited period of up to a few years, to be subsequently released from regular control in accordance with the provisions approved by the regulator and managed by not controlled disposal, use, or download.

*Very low activity waste* — radioactive waste that, even when exceeding the dispensing levels, does not require a high level of isolation and containment, so it can be disposed of in landfill type surface disposal facilities with a reduced regulatory control, in which other hazardous wastes could also be placed.

*Low activity waste* — waste containing limited amounts of long half-life radionuclides. The wastes require a high level of isolation and containment for periods of up to several hundred years and are suitable for final disposal in surface facilities.

*Intermediate activity waste* — wastes that owing to their radionuclide content (particularly those with a long half-life) require a higher degree of isolation and containment than a surface disposal facility can provide. However, these wastes do not require, or may require in a very limited way, that measures be taken during their storage and final disposal to control the release of the heat they generate.

*High level waste* — waste with activity concentrations high enough to generate significant amounts of heat owing to radioactive decay processes, or waste contaminated with large amounts of long half-life radionuclides, such that they need to be considered in the design of the facility planned for its disposal.

Poland
[I–43]

Radioactive waste is classified into three categories with respect to the concentration of radioactive isotopes contained in the waste: low, medium and high level radioactive waste. These categories are further subdivided into subcategories according to the half-lives of radioactive isotopes and the concentration of radioactive isotopes contained in the waste. Liquid waste is additionally classified according to its activity concentration. Spent nuclear fuel intended for disposal is classified as a high level radioactive waste. The low, intermediate and high level waste is subsequently classified into subcategories:

— Transitional waste: waste that will decay within a period of three years to below the value defined in the legislation.
— Short lived waste: waste containing radionuclides with a half-life <30 years with the long lived radionuclide concentration restricted to 4000 kBq/kg in individual waste packages and to an overall average of 400 kBq/kg in the total waste volume.
— Long lived waste: waste whose long lived radionuclides activity exceeds 400 kBq/kg.

Portugal
[I–44]

Uses the IAEA classification system

Republic of
Korea
[I–45]

*High level waste* — radioactive waste whose radioactive concentration and heat generation rate are higher than the levels prescribed by the Regulatory Body:

— Radioactivity concentration >4000 Bq/g for alpha emitting radionuclides with a half-life >20 years
— Heat generation rate > 2 kW/m$^3$

*Low and intermediate level waste* — waste other than HLW. There are three subcategories:

— Very low level radioactive waste
— Low level radioactive waste
— Intermediate level radioactive waste

*Very low level waste*

**Romania**
[I–46]

*Excluded radioactive waste (EW)* — waste containing radionuclides with an activity concentration so small that the waste can be released from regulatory control (conditionally or unconditionally).

*Transitional radioactive waste (TW)* — waste having an activity concentration above clearance levels, but which decays to below clearance levels within a reasonable storage period (not more than five years).

*Very low level radioactive waste (VLLW)* — short lived waste in which the activity concentration is above clearance levels, but with a radioactive content below levels established by the National Commission for Nuclear Activities Control (CNCAN) for defining low level waste. The disposal of very low level waste requires less complex arrangements than the disposal of short lived low level waste.

*Low and intermediate level radioactive waste (LILW)* — radioactive waste in which the activity concentration is above the levels established by CNCAN for the definition of very low level waste, but with a radioactive content and thermal power below those of high level waste. Low level waste does not require shielding during handling or transportation. Intermediate level waste generally requires shielding during handling, but needs little or no provision for heat dissipation during handling or transportation.

— *Long lived radioactive waste:* a waste containing radionuclides with a half-life >30 years in quantities and/or at concentrations of activity above the values established by CNCAN, for which isolation from the biosphere is necessary for more time than the institutional control duration.
— *Short lived radioactive waste:* a radioactive waste that is not long lived.

*High level radioactive waste (HLW) is:*

— Liquid radioactive waste containing the majority of the fission products and actinides existing initially in the spent fuel and forming the residues of the first extraction cycle of reprocessing.
— The solidified radioactive waste of the above and spent fuel.
— Any other radioactive waste with a similar activity concentration range to the waste mentioned above.

**Russian Federation**
[I–47]

*Class 1* — solid high level radioactive waste that will be disposed of in deep disposal facilities after being stored to reduce the decay heat
*Class 2* — solid high level radioactive waste and intermediate level long lived radioactive waste containing radionuclides with half-lives >31 years that will be disposed of in deep disposal facilities without prior storage to reduce the decay heat
*Class 3* — solid intermediate level radioactive waste and low level long lived radioactive waste containing radionuclides with half-lives >31 years that will be disposed of in near surface disposal facilities at a depth of up to 100 m
*Class 4* — solid low level radioactive waste and very low level radioactive waste that will be disposed of in near surface disposal facilities located at ground level
*Class 5* — liquid intermediate level and low level radioactive waste that will be disposed in deep well injection disposal facilities
*Class 6* — radioactive waste generated from mining and processing of uranium ores or during operations that are not associated with atomic energy use, namely, mining and processing of mineral and organic raw materials with increased concentrations of naturally occurring radionuclides that will be disposed of in near surface disposal facilities

**Serbia**
[I–48]

Uses the IAEA classification system

**Slovakia**
[I–49]

Uses the IAEA classification system

Slovenia
[I–50]

*Transitional radioactive waste* — the activity of which decreases during storage below the limit value, enabling their release into environment.

*Very low level radioactive waste* — for which the competent regulatory body for nuclear and radiation safety may approve conditional clearance.

*Low and intermediate level radioactive waste* (LILW) — with insignificant heat generation, classified into two groups:
— *Short lived LILW*: containing radionuclides with a half-life <30 years and specific activity of alpha emitters ≤4000 Bq/g for an individual package, but on average not higher than 400 Bq/g in the overall amount of LILW.
— *Long lived LILW*: where the specific activity of alpha emitters exceeds the limitations for short lived LILW.

*High level radioactive waste* (HLW) — which contains radionuclides whose decay generates such an amount of heat that it has to be considered in its management.

*NORM* — radioactive waste containing naturally occurring radionuclides that are generated in the processing of nuclear mineral materials or other industrial processes and are not sealed sources of radiation in accordance with the regulations on the use of radioactive sources and radiation practices.

South Africa
[I–51]

*High level waste (HLW)* — heat generating radioactive waste with high, long and short lived radionuclide concentrations:
— With thermal power >2 kW/m$^3$
— Long lived alpha, beta and gamma emitting radionuclides at activity concentration levels greater than the levels specified for LILW-LL
— Long lived alpha, beta and gamma emitting radionuclides at activity concentration levels that could result in inherent intrusion dose (the intrusion dose assuming the radioactive waste is spread on the surface) >100 mSv per annum

*Low and intermediate level waste — long lived (LILW-LL)* — radioactive waste with low or intermediate short lived radionuclide and intermediate long lived radionuclide concentrations:
— Thermal power (mainly owing to short lived radio nuclides ($T\frac{1}{2}$ <31 y) <2 kW/m$^3$)

AND
— Long lived alpha radio nuclides ($T\frac{1}{2}$ >31 y) concentrations:
  • Alpha: <4000 Bq/g
  • Beta and gamma: <40 000 Bq/g
OR
— Long lived alpha, beta and gamma emitting radionuclides at activity concentration levels that could result in inherent intrusion dose (the intrusion dose assuming the radioactive waste is spread on the surface) of 10–100 mSv per annum
*Low and intermediate level waste — short lived (LILW-SL)* — radioactive waste with low or intermediate short/lived radionuclide concentrations and/or low long lived radionuclide concentrations:
— Thermal power (mainly owing to short lived radio nuclides ($T\frac{1}{2}$ <31 y) <2 kW/m$^3$)

AND
— Long lived alpha radio nuclides ($T\frac{1}{2}$ >31 y) concentrations:
  • Alpha: <400 Bq/g
  • Beta and gamma: <4000 Bq/g
OR
— Long lived alpha, beta and gamma emitting radionuclides at activity concentration levels that could result in inherent intrusion dose (the intrusion dose assuming the radioactive waste is spread on the surface) <10 mSv per annum

*Very low level waste (VLLW)* — radioactive waste containing a very low concentration of radioactivity

*Naturally occurring radioactive material — low activity (NORM-L)* — potential radioactive waste containing low concentrations of NORM. Long lived radio nuclide concentration <100 Bq/g

*Naturally occurring radioactive materials — enhanced activity (NORM-E)* — radioactive waste containing enhanced concentrations of NORM. Long lived radio nuclide concentration >100 Bq/g

| | |
|---|---|
| Spain<br>[I–52] | *Low and intermediate level wastes (LILW)* — which include those whose activity is mainly owing to the presence of beta or gamma emitting radionuclides with short or intermediate half-lives (<30 years), and whose long lived radionuclide content is very low and limited. This group includes the subcategory of very low level wastes (VLLW).<br><br>*Special waste (SW)* — in accordance with Nuclear Safety Council Instruction IS-29 on safety criteria in temporary storage facilities for spent fuel and high level radioactive waste, this includes the following: nuclear fuel attachments; neutron sources; used in-core instrumentation or substituted components deriving from the reactor vessel system and reactor internals, generally metallic and presenting a high level of radiation through neutron activation; and other waste which, because of its radiological characteristics, is not eligible for management in the existing near surface level definitive disposal facility for LILW in Spain. Its management is connected to that of high level waste.<br><br>*High level wastes (HLW)* — those that contain long lived alpha emitters with half-lives >30 years in appreciable concentrations and that may generate heat as a result of radioactive decay, owing to their high specific activity. This category includes spent fuel. Also included, for the purposes of integral management, are those other intermediate level wastes (ILW) that in view of their characteristics are not eligible for definitive management under the conditions established for El Cabril and that require specific installations for this purpose. |
| Sweden<br>[I–53] | There is no legally defined waste classification system in Sweden for nuclear or radioactive waste. There are, however, established waste acceptance criteria for different disposal routes for nuclear and radioactive waste.<br><br>The waste classification scheme used by the Swedish nuclear industry:<br>— *Cleared material* — material with such small amounts of radioactive nuclides that it has been released from regulatory control.<br>— *Very low level waste short lived (VLLW-SL)* — contains small amounts of short lived nuclides with a half-life <31 years and the dose rate on the waste package is <0.5 mSv/h. Long lived nuclides with a half-life >31 years can be present in restricted quantities.<br>— *Low level waste short lived (LLW-SL)* — contains small amounts of short lived nuclides with a half-life <31 years and the dose rate on the waste package (and unshielded waste) is <2 mSv/h. Long lived nuclides with a half-life >31 years can be present in restricted quantities.<br>— *Intermediate level waste short lived (ILW-SL)* — contains significant amounts of short lived nuclides with a half-life <31 years and the dose rate on the waste package is <500 mSv/h. Long lived nuclides with a half-life >31 years can be present in restricted quantities.<br>— *Low and intermediate long lived waste (LILW-LL)* — contains significant amounts of long lived nuclides with a half-life >31 years, exceeding the restricted quantities for short lived waste.<br>— *High level waste (HLW)* — nuclear fuel; typical decay heat >2 kW/m$^3$ and contains significant amounts of long lived nuclides with a half-life >31 years, exceeding the restricted quantities for short lived wast.e |
| Switzerland<br>[I–54] | *High level waste (HLW)* — vitrified fission product waste from the reprocessing of spent fuel, or spent fuel if declared as waste<br><br>*Alpha toxic waste (ATA)* — waste with a concentration of alpha emitters >20 000 Bq/g of conditioned waste<br><br>*Low and intermediate level waste (L/ILW)* — all other radioactive waste |
| Syrian Arab Republic<br>[I–55] | Uses the IAEA classification system |

**Türkiye**

*Very short lived radioactive waste* — the activities are above exemption levels and after a storage period of a few years at most such waste will be appropriate for clearance. There is no need for disposal; decay storage is the preferred option.

*Very low level rad-waste* — not classified as very short lived waste and has activity concentrations below approximately 100× the clearance limits. Surface disposal is the preferred option for this class of waste.

*Low and intermediate level radioactive waste:*
— After conditioning, alpha emitting radionuclide concentrations are <400 Bq/g for the average of the whole package and <4000 Bq/g for an individual waste package. Near surface disposal is the preferred final solution for this class of waste.
— After conditioning, alpha emitting radionuclide concentrations are >400 Bq/g for the average of the whole package and >4000 Bq/g for an individual waste package. Intermediate depth disposal is the preferred final solution for this class of waste.

*High level waste* — spent fuel declared to be radioactive waste, radioactive wastes that are products of reprocessing and may include fission products and actinides, and other radioactive wastes with comparable activity levels with those above. Deep geological disposal is final solution for this class of waste.

**Ukraine**
**[I–56]**

*Low level waste* — material with the special activities between 10 kBq/kg and 10 MBq/kg
*Intermediate level waste* — material with special activities between 10 MBq/kg and 100 GBq/kg
*High level waste* — materials with special activities >100 GBq/kg

**United Arab Emirates**
**[I–57]**

Uses the IAEA classification system

**United Kingdom**
**[I–58]**

*High level waste (HLW)* — typically the highly active liquor (HAL) by-product from the reprocessing of spent fuel, it is sufficiently radioactive that it generates a significant amount of heat (typically thermal power >~2 kW/m$^3$), which has to be taken into account for its storage and disposal (e.g. in the design of storage and disposal facilities).

*Intermediate level waste (ILW)* — waste with radioactivity levels exceeding the upper boundaries for low level waste (LLW), but which does not require heating to be taken into account in the design of storage or disposal facilities.

*Low level waste (LLW)* — radioactive waste having a radioactive content not exceeding 4 GBq/te of alpha and/or 12 GBq/te of beta/gamma activity.

*Very low level waste (VLLW)* — a subcategory of LLW defined as that which can be safely disposed of alongside municipal, commercial or industrial waste, or at permitted landfill facilities, subject to defined limits on radioactivity content (defined in the specific waste acceptance criteria for each landfill facility).

| | |
|---|---|
| United States of America [I–59] | The USA has two waste classification systems: one for 'civilian' wastes and the other for US DOE (defence related) wastes. For civilian wastes, the categories are based on suitability for near surface disposal: |

*Low level waste (LLW)* — defined in regulations based on suitability for near surface disposal through consideration of concentrations of long and short lived radionuclides. See 10 CFR 61 [USNRC 1982] for full definitions. It is subdivided into:
— *Class A low level waste:* determined by the characteristics listed in 10 CFR 61.55 and physical form requirements in 10 CFR 61.56. (The USA does not have a minimum threshold for class A waste).
— *Class B low level waste:* waste that has to meet more rigorous requirements on waste form than class A waste to ensure stability.
— *Class C low level waste:* waste that not only has to meet more rigorous requirements on waste form than class B waste to ensure stability but also requires additional measures at the disposal facility to protect against inadvertent intrusion.

*Greater than class C waste (GTCC)* — waste that exceeds the limits for class C waste and is not generally acceptable for near surface disposal.

*High level waste (HLW)* — the highly radioactive material resulting from reprocessing of spent fuel, including liquid waste generated directly in reprocessing and any solid material derived from such liquid waste containing fission products in sufficient concentrations and other highly radioactive material that the Nuclear Regulatory Commission (NRC), consistent with existing law, determines by rule requires permanent isolation.

*Spent fuel (SF)* — fuel that has been withdrawn from a nuclear reactor following irradiation, the constituent elements of which have not been separated by reprocessing. For civilian applications, this is considered to be a waste.

*By-product material* (uranium mill tailings) — tailings or wastes generated by the extraction or concentration of uranium or thorium from any ore processed primarily for its source material content (also referred to as AEA 11(e)2 waste).

The US DOE classifies wastes as:
*Low level waste (LLW)* — radioactive waste other than HLW, TRU and by-product material.
*High level waste (HLW)* — similar to civilian definition.
*Transuranic waste (TRU)* — USDOE owned waste (mostly defence related) contaminated with humanmade radioisotopes beyond or 'heavier' than uranium on the periodic table of the elements (long lived alpha emitting waste with concentrations greater than 3700 Bq/g (100 nCi/g)). Subdivided into:
— *Contact handled* TRU (CH): TRU waste with a surface dose rate of <200 millirem per hour
— *Remote handled* TRU (RH): TRU waste with a surface dose rate of 2200 millirem per hour or greater

*By-product material:* similar to civilian definition.
*Spent fuel:* the USDOE does not consider spent fuel to be a waste.

| | |
|---|---|
| Uzbekistan [I–60] | Uses the IAEA classification system |

Viet Nam
[I–61]

*Low level waste, very short lived (LLW-VSL)* — waste contains only very short lived radionuclides (their half-life is <100 days) and can decay to a level lower than clearance levels within five years from generation.

*Low and intermediate level waste, short lived (LILW-SL)* — radioactive waste cannot decay to a level lower than clearance levels within five years from generation and contains radionuclides emitting beta/gamma with a half-life of 100 days to 30 years or contains radionuclides emitting alpha with an average activity concentration ≤400 Bq/g.

*Low and intermediate level waste, long lived (LILW-LL)* — radioactive waste containing radionuclides with a half-life >30 years or radionuclides emitting alpha radiation with an average activity concentration >400 Bq/g but an activity concentration $\leq 10^4$ TBq/m$^3$.

*High level waste (HLW)* — radioactive waste containing radionuclides with activity concentration >$10^4$ TBq/m$^3$.

## REFERENCES TO ANNEX I

[I–1]     ARGENTINE NATIONAL ATOMIC ENERGY COMMISSION, Joint Convention on the Safety of Spent Fuel Management and on the Safety of Radioactive Waste Management — Seventh National Report, (2020),
https://www.iaea.org/sites/default/files/2025-03/argentina-7rm_english.pdf

[I–2]     AUSTRALIAN RADIATION PROTECTION AND NUCLEAR SAFETY AGENCY, Joint Convention on the Safety of Spent Fuel Management and on the Safety of Radioactive Waste Management — National Report of the Commonwealth of Australia (2020),
https://www.arpansa.gov.au/sites/default/files/joint_convention_on_the_safety_of_spent_fuel_management_and_on_the_safety_of_radioactive_waste_management_-_national_report_of_the_commonwealth_of_australia_-_october_2020.pdf

[I–3]     FEDERAL MINISTRY REPUBLIC OF CLIMATE ACTION, ENVIRONMENT, ENERGY, MOBILITY, INNOVATION AND TECHNOLOGY, Seventh National Report of Austria on the Implementation of the Obligations of the Joint Convention on the Safety of Spent Fuel and on the Safety of Radioactive Waste Management (2020),
https://www.iaea.org/sites/default/files/austria-7rm.pdf

[I–4]     DEPARTMENT FOR NUCLEAR AND RADIATION SAFETY OF THE MINISTRY FOR EMERGENCY SITUATIONS OF THE REPUBLIC OF BELARUS, The Seventh National Report of the Republic of Belarus for the Joint Convention on the Safety of Spent Fuel and on the Safety of Radioactive Waste Management, Minsk (2020),
https://inis.iaea.org/records/qbg8e-7c906

[I–5]     KINGDOM OF BELGIUM, Seventh Meeting of the Contracting Parties to the Joint Convention on the Safety of Spent Fuel Management and on the Safety of Radioactive Waste Management (2020),
https://afcn.fgov.be/fr/system/files/2020-11-10-jc-rapport-be-2020.pdf

[I–6]     STATE REGULATORY AGENCY FOR RADIATION AND NUCLEAR SAFETY, Third National Report of Bosnia and Herzegovina on the Implementation of the Obligations under the Joint Convention on the Safety of Spent Fuel and on the Safety of Radioactive Waste Management to the 7th Review Meeting (2020),
https://www.iaea.org/sites/default/files/bosnia-and-herzegovina-7rm.pdf

[I–7]     THE REPUBLIC OF BOTSWANA, The Republic of Botswana Second National Report on Implementation of the Obligations Under the Joint Convention on the Safety of Spent Fuel Management and on the Safety of Radioactive Waste Management (2020),
https://www.iaea.org/sites/default/files/botswana-7rm.pdf

[I–8]     FEDERAL REPUBLIC OF BRAZIL, National Report of Brazil for the Seventh Review Meeting of the Joint Convention on the Safety of Spent Fuel Management and on the Safety of Radioactive Waste Management (2020),
https://www.gov.br/cnen/pt-br/assunto/radioprotecao-e-seguranca-nuclear/copy_of_NationalReport Brazil2020.pdf

[I–9]     THE REPUBLIC OF BULGARIA, Seventh National Report on Fulfilment of the Obligations under the Joint Convention on the Safety of Spent Fuel Management and on the Safety of Radioactive Waste Management (2020),
https://bnra.bg/media/joint-convention-on-the-safety-of-spent-fuel-management-and-on-the-safety-of-radioactive-waste-management-%E2%80%93-7th-report_eng.pdf

[I–10]    CANADIAN NUCLEAR SAFETY COMMISSION, Canadian National Report for the Joint Convention on the Safety of Spent Fuel Management and on the Safety of Radioactive Waste Management — Seventh Report (2020),
https://www.cnsc-ccsn.gc.ca/eng/resources/publications/reports/jointconvention/seventh-report/seventh-report-joint-convention/

[I–11]    CHILEAN NUCLEAR ENERGY COMMISSION, Third National Report of Chile for the Joint Convention on the Safety of Spent Fuel Management and on the Safety of Radioactive Waste Management (2020),
https://www.iaea.org/sites/default/files/2025-03/chile-7rm.pdf

[I–12]    THE PEOPLE'S REPUBLIC OF CHINA, Fifth National Report for the Joint Convention on the Safety of Spent Fuel Management and on the Safety of Radioactive Waste Management (2020),
https://www.iaea.org/sites/default/files/china-7rm_english.pdf

[I–13]    STATE OFFICE FOR NUCLEAR SAFETY OF CROATIA, Seventh National Report on Implementation of the Obligations Under the Joint Convention on the Safety of Spent Fuel Management and on the Safety of Radioactive Waste Management (2020),
https://www.iaea.org/sites/default/files/2025-03/croatia-7rm.pdf

[I–14]    REPUBLIC OF CUBA, Joint Convention on the Safety of Spent Fuel Management and on the Safety of Radioactive Waste Management, Second National Report (2020),
https://www.iaea.org/sites/default/files/cuba-7rm_english.pdf

[I–15]    REPUBLIC OF CYPRUS, National Report on the Implementation of the Obligations under the Joint Convention on the Safety of Spent Fuel Management and on the Safety of Radioactive Waste Management Submitted for the Purposes of the 7th Review Meeting of the Convention (2020).

[I–16]    STATE OFFICE FOR NUCLEAR SAFETY, The Czech Republic National Report under the Joint Convention on the Safety of Spent Fuel Management and on the Safety of Radioactive Waste Management (2020),
https://sujb.gov.cz/fileadmin/sujb/docs/zpravy/narodni_zpravy/NZ_VP_RAO_7_0A.pdf

[I–17]    MINISTRY OF ENVIRONMENT, ENVIRONMENTAL BOARD, Joint Convention on the Safety of Spent Fuel Management and on the Safety of Radioactive Waste Management: Sixth Estonian National Report as Referred to in Article 32 of the Convention (2020).

[I–18]    RADIATION AND NUCLEAR SAFETY AUTHORITY OF FINLAND, Joint Convention on the Safety of Spent Fuel Management and on the Safety of Radioactive Waste Management: 7th Finnish National Report as Referred to in Article 32 of the Convention (2020),
https://www.julkari.fi/bitstream/handle/10024/140601/stuk-b259.pdf?sequence=1&isAllowed=y

[I–19]    GOVERNMENT OF FRANCE, France's Seventh National Report on Compliance with the Joint Convention (2020),
https://www.iaea.org/sites/default/files/france_national_report_7rm.pdf

[I–20]    GOVERNMENT OF GEORGIA, Georgian National Report for the Joint Convention on the Safety of Spent Fuel Management and on the Safety of Radioactive Waste Management. (2020).

[I–21]    FEDERAL MINISTRY FOR THE ENVIRONMENT, NATURE CONSERVATION AND NUCLEAR SAFETY, Joint Convention on the Safety of Spent Fuel Management and on the Safety of Radioactive Waste Management Report of the Federal Republic of Germany for the Seventh Review Meeting in May 2021 (2020),
https://www.bundesumweltministerium.de/en/download/report-of-the-federal-government-for-the-seventh-review-meeting-in-may-2021

[I–22]    REPUBLIC OF GHANA, National Report for the Joint Convention on the Safety of Spent Fuel Management and on the Safety of Radioactive Waste Management, Fourth Report (2020),
https://www.iaea.org/sites/default/files/2025-03/ghana-7rm.pdf

[I–23]    GREEK ATOMIC ENERGY COMMISSION, Joint Convention on the Safety of Spent Fuel Management and on the Safety of Radioactive Waste Management, National Report of Greece (2020),
https://eeae.gr/files/anakoinoseis/reports/joint_convention_national_report_2020.pdf

[I–24]    GOVERNMENT OF HUNGARY, National Report. Seventh Report Prepared within the Framework of the Joint Convention on the Safety of Spent Fuel Management and on the Safety of Radioactive Waste Management (2020).

[I–25]    GOVERNMENT OF THE REPUBLIC OF INDONESIA, National Report on Compliance to Joint Convention on the Safety of Spent Fuel Management and on the Safety of Radioactive Waste Management (2020),
https://www.iaea.org/sites/default/files/2025-03/indonesia-7rm.pdf

[I–26]    INTERNATIONAL ATOMIC ENERGY AGENCY, Classification of Radioactive Waste, IAEA Safety Standards Series No. GS-G-1, IAEA, Vienna (2009).

[I–27]    GOVERNMENT OF ITALY, Joint Convention of the Safety of Spent Fuel Management and on the Safety of Radioactive Waste Management, Sixth Italian National Report (2020).

[I–28]    GOVERNMENT OF JAPAN, National Report of Japan for the Seventh Review Meeting. Joint Convention on the Safety of Spent Fuel Management and on the Safety of Radioactive Waste Management (2020),
https://www.nra.go.jp/english/cooperation/conventions/JC.html

[I–29]    HASHEMITE KINGDOM OF JORDAN, Second National Report for the Joint Convention on the Safety of Spent Fuel Management and on the Safety of Radioactive Waste Management (2020),
https://www.iaea.org/sites/default/files/jordan-7rm.pdf

[I–30]    GOVERNMENT OF THE REPUBLIC OF KAZAKHSTAN, Fourth National Report of the Republic of Kazakhstan on Compliance with the Obligation of the Joint Convention on the Safety of Spent Fuel Management and on the Safety of Radioactive Waste Management (2020).

[I–31]    MINISTRY OF INFRASTRUCTURE AND THE ENVIRONMENT, Joint Convention on the Safety of Spent Fuel Management and on the Safety of Radioactive Waste Management. National Report of the Kingdom of the Netherlands for the Seventh Review Meeting (2020),
https://english.autoriteitnvs.nl/documents/2020/11/03/joint-convention-on-the-safety-of-spent-fuel-management-and-on-the-safety-of-radioactive-waste-management

[I–32]    RADIATION SAFETY CENTRE OF THE STATE ENVIRONMENTAL SERVICE, National Report on the Implementation of the Obligations Under the Convention on the Safety of Spent Fuel Management and on the Safety of Radioactive Waste Management (2020).

[I–33]    STATE NUCLEAR POWER SAFETY INSPECTORATE, Lithuanian National Report Under the Joint Convention on the Safety of Spent Fuel Management and on the Safety of Radioactive Waste Management: Sixth Lithuanian National Report (2020),
https://vatesi.lrv.lt/en/international-cooperation/international-treaties/joint-convention-on-the-safety-of-spent-fuel-management-and-on-the-safety-of-radioactive-waste-management/

[I–34]    INSTITUT NATIONAL DES SCIENCES ET TECHNIQUES NUCLÉAIRES MADAGASCAR INSTN, Joint Convention on the Safety of Spent Fuel Management and on the Safety of Radioactive Waste Management, National Report, Madagascar (2020).

[I–35]    ATOMIC ENERGY REGULATORY AUTHORITY, National Report on the Implementation of the Obligations Under the Joint Convention on the Safety of Spent Fuel Management and on the Safety of Radioactive Waste Management (2022).

[I–36]    RADIATION SAFETY AND NUCLEAR SECURITY AUTHORITY, National Report for Mauritius for the Seventh Review Meeting of Contracting Parties (2020),
https://www.iaea.org/sites/default/files/2025-03/mauritius_national_report_7rm.pdf

[I–37]    GOVERNMENT OF MEXICO, Joint Convention on Safety in the Management of Spent Fuel and on the Safety in the Management of Radioactive Waste, First National Report (2020),
https://www.iaea.org/sites/default/files/2025-03/mexico_national_report_7rm.pdf

[I–38]    GOVERNMENT OF MONTENEGRO, Fourth National Report on the Implementation of Obligations under the Joint Convention on the Safety of Spent Fuel Management and on the Safety of Radioactive Waste Management (2020).

[I–39]    HEAD OF THE GOVERNMENT OF KINGDOM OF MOROCCO, Seventh National Report on the Joint Convention on the Safety of Spent Fuel Management and on the Safety of Radioactive Waste Management (2020),
https://www.iaea.org/sites/default/files/morocco-7rm.pdf

[I–40]   GOVERNMENT OF NIGERIA, Nigerian National Report for Seventh Review Meeting of the Joint Convention on the Safety of Spent Fuel Management and on the Safety of Radioactive Waste Management (2020), https://www.iaea.org/sites/default/files/2025-03/nigeria-7rm.pdf

[I–41]   RADIATION SAFETY DIRECTORATE, Fourth National Report on the Joint Convention on the Safety of Spent Fuel Management and on the Safety of Radioactive Waste Management (2020), https://www.iaea.org/sites/default/files/2025-03/north-macedonia-7rm.pdf

[I–42]   GOVERNMENT OF PARAGUAY, National Report of the Republic of Paraguay According to the Provisions of the Joint Convention on the Safety of Spent Fuel Management and on the Safety of Radioactive Waste Management (2020), https://www.iaea.org/sites/default/files/2025-03/paraguay_national_report_7rm.pdf

[I–43]   NATIONAL ATOMIC ENERGY AGENCY, National Report of Republic of Poland on Compliance with Obligations of the Joint Convention on the Safety of Spent Fuel Management and on the Safety of Radioactive Waste Management. Polish 7th National Report as Referred to in Article 32 of the Joint Convention (2020), https://www.google.com/url?sa=t&rct=j&q=&esrc=s&source=web&cd=&ved=2ahUKEwiSn9r17O6QAxV7S_EDHWrFL0gQFnoECB4QAQ&url=https%3A%2F%2Fwww.gov.pl%2Fattachment%2F907e5d25-2e6b-46ea-8020-e93c7bf9b226&usg=AOvVaw0CerA10rreeRv4pEzdMTWg&opi=89978449

[I–44]   PORTUGUESE ENVIRONMENT AGENCY, Joint Convention on the Safety of Spent Fuel Management and on the Safety of Radioactive Waste Management, Seventh Review Meeting of the Contracting Parties, Fourth National Report by Portugal (2020), https://apambiente.pt/sites/default/files/_Prevencao_gestao_riscos/Protecao_radiologica/DAN_RR/Relat%C3%B3rios/JC_7th_review_report.pdf

[I–45]   NUCLEAR SAFETY AND SECURITY COMMISSION, Korean Seventh National Report under the Joint Convention on Safety of Spent Fuel Management and on the Safety of Radioactive Waste Management (2020), https://www.iaea.org/sites/default/files/korean-rep-of-7rm.pdf

[I–46]   GOVERNMENT OF ROMANIA, Romania, Joint Convention on the Safety of Spent Fuel Management and on the Safety of Radioactive Waste Management, Romanian Seventh National Report (2020), http://www.cncan.ro/assets/Informatii-Publice/06-Rapoarte/2020/Romania-National-ReportJC7th2020.pdf

[I–47]   STATE ATOMIC ENERGY CORPORATION ROSATOM, The Sixth National Report of the Russian Federation on Compliance with the Obligations of the Joint Convention on the Safety of Spent Fuel Management and the Safety of Radioactive Waste Management (2020), https://www.iaea.org/sites/default/files/russian-federation-7rm_english.pdf

[I–48]   SERBIAN RADIATION AND NUCLEAR SAFETY AND SECURITY DIRECTORATE, Second National Report. Joint Convention on the Safety of Spent Fuel Management and on the Safety of Radioactive Waste Management (2020), https://www.srbatom.gov.rs/srbatomm/wp-content/uploads/2023/03/SERBIA-Joint-Convention-Second-National-Report-Corrected.pdf

[I–49]   GOVERNMENT OF THE SLOVAK REPUBLIC, National Report of the Slovak Republic, Compiled in Terms of the Joint Convention on the Safety of Spent Fuel Management and on the Safety of Radwaste Management (2020), https://www.ujd.gov.sk/wp-content/uploads/2021/11/JC_NR_2020_ang.pdf

[I–50]   SLOVENIAN NUCLEAR SAFETY ADMINISTRATION, Seventh Slovenian Report Under the Joint Convention on the Safety of Spent Fuel Management and on the Safety of Radioactive Waste Management (2020), https://www.gov.si/en/news/2020-11-03-7th-national-report-under-the-joint-convention-on-the-safety-of-spent-fuel-management-and-on-the-safety-of-radioactive-waste-management/

[I–51]   SOUTH AFRICAN NATIONAL NUCLEAR REGULATOR, South African National Report on the Compliance to Obligations Under the Joint Convention on Safety of Spent Fuel Management and on the Safety of Radioactive Waste Management (2020), https://nnr.co.za/wp-content/uploads/2023/02/SOUTH-AFRICAN-NATIONAL-REPORT-ON-COMPLIANCE-TO-OBLIGATIONS-UNDER-THE-JOINT-CONVENTION-SUBMISSION-TO-THE-SEVENTH-REVIEW-MEETING-OF-THE-CONTRACTING-PARTIES-OCTOBER-2020.pdf

[I–52]   GOVERNMENT OF SPAIN, Joint Convention on the Safety of Spent Fuel Management and on the Safety of Radioactive Waste Management. Seventh National Report (2020), https://www.miteco.gob.es/content/dam/miteco/es/energia/files-1/nuclear/Residuos/GestionResiduos/Convencion/DocConvencion7/Seventh_Spanish_National_Report.pdf

[I–53]  MINISTRY OF THE ENVIRONMENT, Sweden's Seventh National Report Under the Joint Convention on the Safety of Spent Fuel Management and on the Safety of Radioactive Waste Management. Sweden's Implementation of the Obligations of the Joint Convention (2020),
https://www.regeringen.se/rattsliga-dokument/departementsserien-och-promemorior/2020/10/ds-202021/

[I–54]  SWISS FEDERAL NUCLEAR SAFETY INSPECTORATE ENSI, Implementation of the Obligations of the Joint Convention on the Safety of Spent Fuel Management and on the Safety of Radioactive Waste Management. 7th National Report of Switzerland in Accordance with Article 32 of the Convention (2020),
https://ensi.admin.ch/de/wp-content/uploads/sites/2/2020/10/Joint_Convention-Seventh_national_report-Switzerland_2020_web.pdf

[I–55]  ATOMIC ENERGY COMMISSION OF SYRIA, National Report on the Implementation of the Obligations Under the Joint Convention on the Safety of Spent Fuel Management and on the Safety of Radioactive Waste Management (2022),
https://www.iaea.org/sites/default/files/2025-03/syria_national_report_7rm.pdf

[I–56]  STATE NUCLEAR REGULATORY INSPECTORATE OF UKRAINE, Ukraine National Report on Compliance with Obligations Under the Joint Convention on the Safety of Spent Fuel Management and on the Safety of Radioactive Waste Management (2020),
https://www.iaea.org/sites/default/files/ukraine-7rm.pdf

[I–57]  GOVERNMENT OF UNITED ARAB EMIRATES, United Arab Emirates Fourth National Report on Compliance with the Obligations of the Joint Convention on the Safety Spent Fuel Management and on the Safety of Radioactive Waste Management (2020),
https://www.iaea.org/sites/default/files/2025-03/uae-7rm.pdf

[I–58]  DEPARTMENT FOR BUSINESS, ENERGY AND INDUSTRIAL STRATEGY, The United Kingdom's Sixth National Report on Compliance with the Obligations of the Joint Convention on the Safety of Spent Fuel and Radioactive Waste Management (2020),
https://www.gov.uk/government/publications/the-uks-sixth-national-report-on-compliance-with-the-obligations-of-the-joint-convention-on-the-safety-of-spent-fuel-and-radioactive-waste-management

[I–59]  UNITED STATES DEPARTMENT OF ENERGY, United States of America National Report for the Seventh Review Meeting of the Joint Convention on the Safety of Spent Fuel Management and on the Safety of Radioactive Waste Management (2020),
https://www.energy.gov/sites/prod/files/2020/10/f80/7th-JC-RM-United-States-NR-Final-Oct-2020.pdf

[I–60]  GOVERNMENT OF UZBEKISTAN, The Second National Report of the Republic of Uzbekistan on Compliance with the Obligations of the Joint Convention on the Safety of Spent Fuel Management and on the Safety of Radioactive Waste Management (2020),
https://www.iaea.org/sites/default/files/2025-03/uzbekistan-7rm_english.pdf

[I–61]  GOVERNMENT OF VIET NAM, The Socialist Republic of Viet Nam, Third National Report on Implementation of the Obligations Under the Joint Convention on the Safety of Spent Fuel Management and on the Safety of Radioactive Waste Management (2020).

# Annex II

# STATUS OF NATIONAL LONG TERM MANAGEMENT POLICIES

| Country | Non-nuclear fuel cycle waste | Disused sealed sources | Nuclear fuel cycle waste | Spent fuel (SF) |
|---|---|---|---|---|
| Albania | Storage | Return to supplier or storage | n.a.[a] | n.a. |
| Argentina | Disposal in existing and planned LLW disposal facilities. It is planned to dispose of ILW in DGR | Storage awaiting disposal | Disposal in existing and planned LLW disposal facilities. ILW in storage and will be disposed of in planned DGR | Reprocessing decision deferred, a tentative schedule for disposal will be established together with a strategic plan |
| Armenia | Planned disposal of VLLW, LLW and short lived ILW in near surface disposal facility. Long lived ILW and HLW in DGR | Return to supplier for Category 1, 2 and 3 For other DSRS categories, long-term management policy is awaiting the creation of WMO | Planned disposal of VLLW, LLW and short lived ILW in near surface disposal facility. Long lived ILW and HLW in DGR | Dry storage |
| Australia | Planned disposal in facilities for LLW and ILW | Return to supplier where possible or storage awaiting disposal | n.a. | RR[b] fuel sent for reprocessing, the resulting ILW stored awaiting disposal |
| Austria | Storage awaiting disposal | Return to supplier or storage | n.a. | Return of RR fuel to country of origin |
| Belarus | Long term storage and then disposal in planned disposal facility for VLLW, LLW | Return to supplier or long term storage | For VLLW and LLW the same management as for application waste. For ILW and HLW disposal in DGR after long term storage | Return to Russian Federation for processing |
| Belgium | Currently stored. It is planned to dispose of LLW in a near surface facility | Return to supplier or storage | Currently stored at central facility. Near surface disposal planned for LLW in Dessel and geological disposal still to be confirmed | Long term policy still to be defined — disposal of waste from reprocessing or direct disposal |
| Bosnia and Herzegovina | Storage | Return to the supplier | n.a. | n.a. |
| Botswana | Storage | Return to the supplier | n.a. | n.a. |

| Country (cont.) | Non-nuclear fuel cycle waste | Disused sealed sources | Nuclear fuel cycle waste | Spent fuel (SF) |
|---|---|---|---|---|
| Brazil | Storage awaiting disposal | Return to supplier or storage | Not defined | No decision. Current policy is storage at reactor site pending a decision |
| Bulgaria | Disposal. Near surface disposal facility for LLW in construction. For ILW and HLW long term storage | Return to the supplier or disposal | Disposal. Near surface disposal facility for LLW under construction. For ILW and HLW long term storage | Reprocessing abroad |
| Canada | Storage. Disposal facility for VLLW/ LLW in process | Return to supplier, reuse or storage | Existing storage by each major waste owner. Disposal facility for VLLW/ LLW in process | Planned deep disposal at a volunteer host site in either crystalline or sedimentary rock |
| Chile | Storage | Return to supplier or storage | n.a. | Long term dry storage for SF from RR |
| China | Storage and disposal according to waste category | Return to supplier or storage | Planned geological disposal for HLW and disposal of LLW in near surface disposal facilities (existing and planned) | Reprocessing of nuclear power plant spent fuel (NPP SF), direct disposal of certain types of SF is not ruled out. Research reactor SF, planned deep disposal co-located with HLW |
| Croatia | Long term storage followed by disposal | Return to supplier or storage | Long term storage followed by near surface disposal | Dry storage followed by disposal |
| Cuba | Storage | Return to supplier or storage | n.a. | n.a. |
| Cyprus | Decay storage | Return to the supplier | n.a. | n.a. |
| Czech Republic | Disposal in operating disposal facilities and in planned DGR | Disused sealed sources are disposed of in operating disposal facilities and in planned DGR or returned to the country of origin | Existing surface disposal (at Dukovany power plant site) Disposal in operating disposal facilities and in planned DGR | Direct disposal in DGR is preferred option, but other options are not excluded |
| Denmark | Storage awaiting disposal | Return to supplier or storage | n.a. | Dual track — either disposal in Denmark or international solution |
| Estonia | Storage awaiting disposal | Return to supplier or storage awaiting disposal | n.a. | n.a. |

| Country (cont.) | Non-nuclear fuel cycle waste | Disused sealed sources | Nuclear fuel cycle waste | Spent fuel (SF) |
| --- | --- | --- | --- | --- |
| Finland | Disposal in silos | Return to supplier or disposal | Disposal in intermediate depth bedrock, existing underground cavern disposal (at each reactor site) | Disposal in bedrock. Construction of the encapsulation plant and the disposal facility started in 2016 |
| France | Existing disposal options for VLLW and LLW | Return to supplier. Disposal or recycling routes being implemented | Existing disposal options for VLLW and LLW. Storage for the other waste | Reprocessing then disposal of resulting waste |
| Georgia | Storage awaiting disposal | Storage awaiting disposal | n.a. | n.a. |
| Germany | Storage awaiting disposal | Return to supplier or storage awaiting disposal | Storage with the objective of disposal in deep geological formations | Planned deep disposal for NPP SF, site not yet decided. For RR fuels, return to country of origin, or manage with NPP fuel |
| Ghana | Return to supplier or storage at centralized waste processing facility | Return to supplier or storage | n.a. | n.a. |
| Greece | Stored in RW Interim Storage and Management Facility until final management solution | Return to supplier | n.a. | n.a. |
| Hungary | Disposal in existing near surface repository at Püspökszilágy or deep geological disposal facility | Disposal | Disposal at the National Radioactive Waste Repository in Bátaapáti, or deep geological disposal | Deep geological disposal for SF from NPP. For RR SF repatriation to the manufacturer's country |
| Iceland | Return to supplier | Return to supplier | n.a. | n.a. |
| Indonesia | Returned to supplier of managed by Radioactive Waste Technology Centre | Return to supplier or storage | Stored until possible disposal | Return of RR fuel to the country of origin |
| Iran, Islamic Republic of | Treatment and storage | Storage and reuse | Storage until possible disposal | n.a. |
| Ireland | Storage | Return to supplier or overseas for recycling/ reuse | n.a. | n.a. |

| Country (cont.) | Non-nuclear fuel cycle waste | Disused sealed sources | Nuclear fuel cycle waste | Spent fuel (SF) |
|---|---|---|---|---|
| Italy | Central storage. Planned national LLW and ILW near surface disposal facility | Return to supplier or storage | Planned VLLW/LLW near surface disposal facility and storage facility for ILW/HLW | Reprocessing abroad, long term storage of remaining SF |
| Japan | Near surface disposal | Return to supplier or storage | Planned geological disposal, intermediate depth disposal and near surface disposal. There are existing LLW disposal facilities | Reprocessing |
| Jordan | Long term storage until disposal in near surface LILW disposal site | Return to the supplier if possible or storage | Long term storage | Return to country of origin |
| Kazakhstan | Long term storage | Storage | Long term storage | Long term storage |
| Korea, Republic of | Engineered shallow land disposal facility is under review | Storage or disposal | Engineered shallow land disposal facility under review | On-site storage until operation of an interim storage facility and development of HLW disposal facility |
| Latvia | Disposal of LLW and storage of ILW | Return to supplier or storage pending disposal | n.a. | n.a. |
| Lithuania | Storage awaiting disposal | Return to supplier or storage | Storage awaiting disposal | Stored for 50 years. Disposal planned in DGR |
| Luxembourg | Export to Belgium | Return to supplier or export to Belgium | n.a. | n.a. |
| Madagascar | Storage | Storage | n.a. | n.a. |
| Malta | Export when possible or storage | Return to supplier or storage | n.a. | n.a. |
| Mexico | Storage awaiting disposal | Return to supplier or storage | Storage awaiting disposal | Disposal in planned DGR for both NPP and RR SF |
| Montenegro | Storage | Return to supplier or storage | n.a. | n.a. |
| Morocco | Storage | Return to supplier or storage | n.a. | Wet storage |

| Country (cont.) | Non-nuclear fuel cycle waste | Disused sealed sources | Nuclear fuel cycle waste | Spent fuel (SF) |
| --- | --- | --- | --- | --- |
| Netherlands, Kingdom of the | Storage (at COVRA) followed by future free release or disposal | Return to supplier or storage | Existing 100 year storage (at COVRA), followed by planned deep disposal for all waste types in a single facility | Reprocessing Existing 100 year storage (at COVRA), followed by planned deep disposal for all waste types in a single facility |
| Nigeria | Storage | Return to supplier or storage | Planned front end activities | Return of SF from RR to supplier or storage |
| North Macedonia | | Return to supplier or storage | n.a. | n.a. |
| Norway | Disposal of LLW at KLDRA Himdalen | Return to supplier or disposal | Disposal of LLW at KLDRA Himdalen. New facility needed for higher activity waste | RR is currently stored with ongoing assessment for management options |
| Oman | Storage | Return to supplier | n.a. | n.a. |
| Peru | Storage | Return to supplier or storage | Stored until possible to send to the radioactive waste management plant | Storage in spent fuel pool at RR site until repatriation to the supplier is possible |
| Poland | Disposal in Rozan Repository or in planned near surface repository | Return to supplier or disposal | Disposal in planned near surface repository and in planned DGR | RR spent fuel is transported to the country of origin |
| Portugal | Storage | Return to supplier or storage | n.a. | RR SF returned to the USA |
| Republic of Moldova | Storage until possible disposal | Return to supplier or storage pending disposal | n.a. | n.a. |
| Romania | Near surface disposal for short lived waste in construction. Long lived waste will be disposed of in DGR | Return to supplier or storage | Near surface disposal for short lived waste in construction. Long lived waste will be disposed of in DGR | Storage and then disposal in planned DGR. Return to the supplier in case of SF of RR |
| Russian Federation | Processing and storage awaiting disposal | Treatment and disposal | Processing and transfer for disposal. There are near surface facilities for LLW and ILW | Storage and reprocessing |
| Saudi Arabia | Site is chosen for central waste management facility | Return to supplier or storage | n.a. | n.a. |

| Country (cont.) | Non-nuclear fuel cycle waste | Disused sealed sources | Nuclear fuel cycle waste | Spent fuel (SF) |
| --- | --- | --- | --- | --- |
| Senegal | Storage | Return to supplier | n.a. | n.a. |
| Serbia | Storage awaiting disposal | Return to supplier or storage | n.a. | n.a. |
| Slovakia | Disposal in existing (Mochovce site) and planned disposal facility | Return to supplier or storage | Existing surface disposal at Mochovce | Long term wet and dry storage. Looking at geological repository or multinational solution |
| Slovenia | Central Storage and then disposal | Return to supplier or storage | Planned near surface and deep geological disposal facilities | Geological disposal is reference scenario; however, the multinational option is being kept open |
| South Africa | Existing near surface disposal at Vaalputs. Possible medium depth disposal facility for long lived waste | Return to supplier or storage | Existing near surface disposal at Vaalputs. Possible medium depth disposal facility for long lived waste | Long term storage with possible reprocessing or disposal in DGR |
| Spain | Existing surface disposal (at El Cabril) for VLLW and LLW | Return to supplier or storage | Existing surface disposal (at El Cabril) for VLLW and LLW. Waste from SF reprocessing will be disposed of together with SF | Storage for 60 years at centralized storage or upgraded storage facilities until DGR is available |
| Sweden | Disposal in nuclear fuel cycle facilities when appropriate | Return to supplier or stored/disposed of | Existing LLW disposal facility, planned long lived waste disposal facility, as well as geological disposal | Planned DGR at Forsmark site |
| Switzerland | Disposal in planned DGR for L/ILW | Recycling if possible, otherwise management as radioactive waste | Planned deep disposal in DGR, co-located repository for long lived wastes and HLW | Deep geological disposal in planned DGR. Ban on reprocessing |
| Syrian Arab Republic | Storage awaiting disposal | Return to supplier or storage | n.a. | n.a. |
| Thailand | Storage | Return to supplier or storage | n.a. | Return of RR SF to country of origin |
| Türkiye | Storage | Return to supplier or storage | Storage and disposal | Long term storage at NPP site |
| Ukraine | Storage and disposal at Vektor site | Return to supplier or storage | Storage and disposal at Vektor site | Awaiting decision, centralized storage |

| Country (cont.) | Non-nuclear<br>fuel cycle waste | Disused<br>sealed sources | Nuclear fuel<br>cycle waste | Spent<br>fuel (SF) |
| --- | --- | --- | --- | --- |
| United Arab Emirates | Storage | Return to the supplier | Disposal in planned near surface repository for LLW or geological disposal facility for ILW and HLW | Reference scenario is disposal in DGR; other options are still under review |
| United Kingdom | Existing surface disposal (at LLWR & Dounreay). Conventional surface landfill facilities for management of VLLW. Other facilities may be developed if required | Return to supplier or storage | Existing surface disposal (at LLWR & Dounreay). Other facilities may be developed if required | Disposal of SF in DGR |
| United States of America | For LLW near surface disposal. For ILW the disposal path to be determined | Return to supplier. Disposal, reuse or recycle | For HLW disposal in geological repository. For LLW near surface disposal | Disposal in geological repository after storage in wet or dry storages |
| Uruguay | Return to the producer or storage | Return to supplier or storage | n.a. | n.a. |
| Uzbekistan | Storage | Return to supplier or storage | Storage | Return of SF from RR to supplier |
| Viet Nam | Central storage, awaiting disposal | Return to supplier or storage | Central storage, awaiting disposal | Return of SF from RR to supplier |
| Zimbabwe | Storage | Return to the supplier | n.a. | n.a. |

[a] n.a.: not applicable.

[b] RR: research reactor.

# NATURE AND ROLE OF THE WASTE MANAGEMENT ORGANIZATION (AS OF 31 DECEMBER 2019)

| Country | Waste management organization (WMO) | Responsibilities | Ownership |
|---|---|---|---|
| Albania | Institute of Applied Nuclear Physics | Management and storage of radioactive waste | State |
| Argentina | National Radioactive Waste Management Program — Argentine Atomic Energy Commission (PNGRR - CNEA) | Management of radioactive waste and surveillance of spent fuel | State |
| Armenia | No specified WMO | n.a.[a] | n.a.[a] |
| Australia | Australian Radioactive Waste Agency (ARWA) | Management of radioactive waste | State |
| Austria | Nuclear Engineering Seibersdorf GmBH (NES) | Predisposal management of radioactive waste | State |
| Azerbaijan | Specialized Enterprise 'Isotope' | Management of radioactive waste | State |
| Belarus | Specialized Enterprise UE Ekores | Management of radioactive waste | State |
| Belgium | ONDRAF/NIRAS | Development and operation of disposal facilities for all types of radioactive waste and spent fuel | State |
| Plurinational State of Bolivia | Bolivian Nuclear Energy Agency (ABEN) | Management of radioactive waste | State |
| Bosnia and Herzegovina | No specified WMO | n.a. | n.a. |
| Bulgaria | State Enterprise Radioactive Wastes (SE RAW) | Management of radioactive waste | State |
| Canada | Nuclear Waste Management Organization (NWMO) | Development and operation of disposal facility for spent fuel | Utilities |
| | Low Level Radioactive Waste Management Office (LLRWMO) | Cleanup and management of Canada's historic waste | State/private |
| | (Other waste owners) | Management and disposal of their own wastes | Utilities/State/private |

| Country (cont.) | Waste management organization (WMO) | Responsibilities | Ownership |
| --- | --- | --- | --- |
| Chile | Chilean Nuclear Energy Commission | Management of radioactive waste | State |
| China | No specified WMO | n.a. | n.a. |
| Croatia | Fund for the Financing of the Decommissioning and Disposal of Radioactive Waste and Spent Nuclear Fuel from the Krško Nuclear Power Plant | Management of radioactive waste | State |
| Cuba | Centre for Radiation Protection and Hygiene (CPHR) | Management of radioactive waste | State |
| Cyprus | No specified WMO | n.a. | n.a. |
| Czech Republic | SÚRAO | Development and operation of radioactive waste and spent fuel storage and disposal facilities | State |
| Denmark | Danish Decommissioning | Management of all radioactive waste | State |
| Estonia | A.L.A.R.A. Ltd | Management of all radioactive waste | State |
| Finland | Posiva Oy | Development and operation of disposal facility for spent fuel. Low level waste disposal is the direct responsibility of the nuclear power plants (NPPs) | Utilities |
| France | ANDRA | Development and operation of disposal facilities for all types of radioactive waste | State |
| Georgia | Department for Radioactive Waste Management | Management of radioactive waste | State |
| Germany | BGZ | Predisposal management of radioactive waste | State |
|  | BGE | Disposal of radioactive waste | State |
| Ghana | Radioactive Waste Management Centre (RWMC) | Management of radioactive waste | State |
| Greece | National Centre of Scientific Research 'Demokritos' | Management of radioactive waste | State |

| Country (cont.) | Waste management organization (WMO) | Responsibilities | Ownership |
|---|---|---|---|
| Hungary | PURAM | Development and operation of storage and disposal facilities for all types of radioactive waste and spent fuel and decommissioning of nuclear facilities | State |
| Iceland | No specified WMO | n.a. | n.a. |
| Indonesia | National Nuclear Energy Agency (BATAN) | Management of spent fuel and radioactive waste | State |
| Iran, Islamic Republic of | Iran Radioactive Waste Management Company (IRWA) | Management of radioactive waste | State |
| Ireland | No specified WMO | n.a. | n.a. |
| Italy | SOGIN | Decommissioning of nuclear facilities and management of radioactive waste | State |
| Japan | Nuclear Waste Management Organization (NUMO) | Development and operation of disposal facility for HLW | State |
| Jordan | Jordan Atomic Energy Commission (JAEC) | Management of spent fuel and radioactive waste | State |
| Korea, Republic of | Korea Radioactive Waste Management Corporation (KORAD) | Development and operation of storage and disposal facilities for all types of radioactive waste and spent fuel, and management of radioactive waste management fund | State |
| Latvia | State Ltd Latvian Environment, Geology and Meteorology Centre | Management of all radioactive waste | State |
| Lithuania | Ignalina NPP | Management of radioactive waste and spent fuel | State |
| Luxembourg | No specified WMO | n.a. | n.a. |
| Malaysia | Nuclear Malaysia | Management of radioactive waste | State |
| Mexico | National Institute of Nuclear Research (ININ) | Management of radioactive waste | State |
| | Laguna Verde NPP | Management of radioactive waste and spent fuel | State |
| Republic of Moldova | Radioactive Waste Storage Facility (RWSF) | Storage of radioactive waste | State |

| Country (cont.) | Waste management organization (WMO) | Responsibilities | Ownership |
|---|---|---|---|
| Morocco | National Centre for Nuclear Energy, Science and Technology (CNESTEN) | Management of radioactive waste | State |
| Netherlands, Kingdom of the | COVRA | Management of radioactive waste | State |
| North Macedonia | No specified WMO | n.a. | n.a. |
| Norway | Norwegian Nuclear Decommissioning (NND) | Decommissioning and management of radioactive waste | State |
| Paraguay | No specified WMO | n.a. | n.a. |
| Peru | RASCO Nuclear Centre | Management of radioactive waste | State |
| Poland | Radioactive Waste Management Plant (ZUOP) | Management of radioactive waste and spent fuel | State |
| Portugal | School of Engineering of the University of Lisbon (IST) | Management of radioactive waste | State |
| Romania | Nuclear and Radioactive Waste Agency (ANDR) | Development and operation of disposal facilities for all types of radioactive waste and spent fuel | State |
| Russian Federation | Federal State Unitary Enterprise National Operator for Radioactive Waste Management (NO RAO) | Development and operation of disposal facilities for all types of radioactive waste. Predisposal management is distributed among several organizations; FSUE RosRAO, Moscow Radon, etc. | State |
| Saudi Arabia | King Abdullah City for Atomic and Renewable Energy | Management of radioactive waste | State |
| Serbia | Public Company Nuclear Facilities of Serbia | Management of radioactive waste | State |
| Slovakia | Nuclear Decommissioning Company (JAVYS) | Development and operation of storage and disposal facilities for all types of radioactive waste and spent fuel, operation of centralized waste processing facilities and decommissioning of nuclear facilities | State |

| Country (cont.) | Waste management organization (WMO) | Responsibilities | Ownership |
|---|---|---|---|
| Slovenia | Agency for Radwaste Management (ARAO) | Development and operation of storage and disposal facilities for all types of radioactive waste and spent fuel | State |
| South Africa | National Radioactive Waste Disposal Institute (NRWDI) | Management of radioactive waste and spent fuel | State |
| Spain | ENRESA | Development and operation of storage and disposal facilities for all types of radioactive waste and spent fuel. Decommissioning of reactors | State |
| Sweden | SKB | Development and operation of storage and disposal facilities for all types of radioactive waste and spent fuel | Utilities |
| Türkiye | TENMAK | Management of radioactive waste | State |
| Ukraine | SA Radon | Management of radioactive waste | State |
| United Kingdom | NDA | Overseeing strategic management of radioactive waste and spent fuel, including waste from historic operations | State |
| United States of America | USDOE | Development and operation of disposal facilities for all spent fuel, certain ILW (greater than class C LLW), and US DOE owned or generated radioactive waste | State |
| | States/compacts | Responsible for disposal of LLW (disposal occurs at commercially operated facilities) | |
| Viet Nam | No specified WMO | n.a. | n.a. |

[a] n.a.: not applicable.

**Annex IV**

**FINANCING SCHEMES AND FUNDING MECHANISMS
FOR SPENT FUEL AND RADIOACTIVE WASTE**

| Country | Funding of radioactive waste management (RWM) | Funding of spent fuel (SF) and high level waste (HLW) management | Funding of decommissioning |
|---|---|---|---|
| Albania | Producers pay for RWM | n.a. | n.a. |
| Argentina | Producers pay for RWM. Governmental funding if produced in the State owned facility | Governmental funding | Governmental funding when the facility is state owned, otherwise the owner of facility |
| Armenia | Producers pay for RWM in case of waste from the nuclear power plant (NPP), governmental funding for other institutional waste | Facility funding | Decommissioning fund |
| Australia | Governmental funding | Governmental funding | Governmental funding |
| Austria | Producers pay for RWM and governmental funding | Governmental funding | Governmental funding |
| Belarus | Operator's financial assets or state budget | Operator's financial assets | Operator's financial assets or state budget |
| Belgium | Producer pays, contribution to ONDRAF/NIRAS long term fund | NPP operators contribute to the fund | NPP operators contribute to the fund; various funds for historical liabilities fed by state |
| Bosnia and Herzegovina | Producer pays or governmental funding | n.a. | Licensee funding |
| Botswana | Producer pays or governmental funding | n.a. | n.a. |
| Brazil | Operator or governmental funding | Operator (governmental) funding | Operator (governmental) funding |
| Bulgaria | Payments to radioactive waste fund | Operators payments to radioactive waste fund and international contributors | Operators payments to Nuclear Facilities Decommissioning Fund and international contributors |
| Canada | Producers pay for RWM | Each licensee has to create its fund | Each licensee has to create its fund |
| Chile | Producers pay for RWM | Governmental funding | Governmental funding |

| Country (cont.) | Funding of radioactive waste management (RWM) | Funding of spent fuel (SF) and high level waste (HLW) management | Funding of decommissioning |
| --- | --- | --- | --- |
| China | Producers pay for RWM | Collection of the funds into a dedicated account based the electricity production | Provided by the generator and the related government |
| Croatia | Producers pay for RWM | The Fund for Financing the Decommissioning of the Krsko Nuclear Power Plant and the Disposal of Radioactive Waste and Spent Nuclear Fuel | The Fund for Financing the Decommissioning of the Krsko Nuclear Power Plant and the Disposal of Radioactive Waste and Spent Nuclear Fuel |
| Cuba | Producers pay for RWM | n.a. | Producer's responsibility |
| Cyprus | Producers pay for RWM | n.a. | n.a. |
| Czech Republic | Producers pay for RWM to specific fund held by Government | Specific fund by licence holders held by Government | Decommissioning fund |
| Denmark | Producers pay for RWM | Governmental funding | Governmental funding |
| Estonia | Producers pay for RWM | The State carries the financial liability | Governmental funding |
| Finland | Producers pay for RWM | Nuclear Waste Management Fund | Nuclear Waste Management Fund |
| France | Producers pay for RWM | The licensee finances and dedicated assets are required | The licensee finances and dedicated assets are required |
| Georgia | Producers pay for RWM. In case of legacy waste founded by State | n.a. | Governmental funding |
| Germany | Producers pay for RWM. Private facilities setting aside provisions. State owned facilities financed by public funds. Small waste producers pay fees to the Land collecting facilities | Private facilities setting aside provisions. State owned facilities financed by public funds | Private facilities setting aside provisions. State owned facilities financed by public funds |
| Ghana | Producers pay for RWM | Governmental funding | Producer's responsibility |
| Greece | Producers pay for RWM | n.a. | Licensee, governmental funding |
| Hungary | Central Nuclear Financial Fund | Central Nuclear Financial Fund | Central Nuclear Financial Fund |
| Iceland | Producers pay for RWM | n.a. | n.a. |
| Indonesia | Producers pay for RWM | Producer's responsibility | Producer's responsibility |
| Ireland | Producers pay for RWM | n.a. | n.a. |

| Country (cont.) | Funding of radioactive waste management (RWM) | Funding of spent fuel (SF) and high level waste (HLW) management | Funding of decommissioning |
| --- | --- | --- | --- |
| Italy | Producers pay for RWM | Partly funds set aside by NPP, but owing to early shutdown, these are insufficient. Additionally, levy on electricity | Governmental funding |
| Japan | Producers pay for RWM | Licensees pay into the reserve fund | Licensees pay into the reserve fund |
| Jordan | Producers pay for RWM | The State carries the financial liability | Governmental funding |
| Kazakhstan | Producers pay for RWM | Governmental funding and international donors | Governmental funding and international donors |
| Korea, Republic of | Producers pay for RWM | Radioactive waste management fund operated by Government (KORAD) | Decommissioning cost of NPPs is covered by Korea Hydro & Nuclear Power Co. and for research reactors by the Government |
| Latvia | Producers pay for RW predisposal management, State pays for disposal | n.a. | Governmental funding |
| Lithuania | Producers pay for RWM | Funds provided by NPP, State and international contributors | Funds provided by NPP, State and international contributors |
| Lesotho | Producers pay for RWM | n.a. | n.a. |
| Luxembourg | Producers pay for RWM | n.a. | n.a. |
| Madagascar | n.a. | n.a. | n.a. |
| Malaysia | Producers pay for RWM | Producer's responsibility | Producer's responsibility |
| Malta | Producers pay for RWM | n.a. | n.a. |
| Mexico | Producers pay for RWM, governmental funding | Producer's responsibility, governmental funding | Producer's responsibility |
| Morocco | Producers pay for RWM | n.a. | Governmental funding |
| Netherlands, Kingdom of the | Producers pay for RWM | Producers fund the processing and long term management | Producer's responsibility |
| Nigeria | Producers pay for RWM | Producer's responsibility | Producer's responsibility |
| North Macedonia | Producers pay for RWM | n.a. | n.a. |
| Norway | Producers pay for RWM | Producer's responsibility | Producers pay, partly governmental funding |

| Country (cont.) | Funding of radioactive waste management (RWM) | Funding of spent fuel (SF) and high level waste (HLW) management | Funding of decommissioning |
| --- | --- | --- | --- |
| Oman | Producers pay for RWM | n.a. | Producers pay |
| Peru | Producers pay for RWM | Producer's (State or private) responsibility | Producers pay |
| Poland | Producers pay for RWM | Decommissioning fund or State budget (in case of the research reactor) | Decommissioning fund or State budget (in case of the research reactor) |
| Portugal | Producers pay for RWM, partly covered by governmental funds | Governmental funds | Governmental funds |
| Republic of Moldova | Producers pay for RWM | n.a. | Governmental funds |
| Romania | Producers pay for RWM | Producer has to pay fee to radioactive waste management funds | Producer has to pay fee to Decommissioning Fund or State Budget (in case of the research reactor) |
| Russian Federation | Producers pay for RWM | Fund contributed to by operators and Government | Fund contributed to by operators and Government |
| Saudi Arabia | Producers pay for RWM | n.a. | National fund |
| Senegal | Producers pay for RWM | Producer's responsibility | Producer's responsibility |
| Serbia | Producers pay for RWM or governmental funding | Governmental funding | Governmental funding |
| Slovakia | Producers pay for RWM | Producer's responsibility, National Nuclear Fund | National Nuclear Fund paid into by operators and State; in case of NPP V-1 the EU also contributes |
| Slovenia | Producers pay for RWM | Fund raised by NPP — Decommissioning Fund | Fund raised by NPP — Decommissioning Fund |
| South Africa | Producers pay for RWM | Producer's responsibility | Owners/waste producers pay |
| Spain | Producers pay for RWM | Fund from NPP operators and payments for waste management services | Fund from NPP operators and payments for waste management services |
| Sweden | Producers pay for RWM | Nuclear Waste Fund — collected as a fee on nuclear power production | Nuclear Waste Fund included in the fee for SF and decommissioning |
| Switzerland | Producers pay for RWM | Liability is with the NPP owners, after NPP shutdown. Waste Management Fund | Liability is with the NPP owners, after NPP shutdown. Decommissioning Fund |
| Syrian Arab Republic | Producers pay for RWM | n.a. | n.a. |

| Country (cont.) | Funding of radioactive waste management (RWM) | Funding of spent fuel (SF) and high level waste (HLW) management | Funding of decommissioning |
| --- | --- | --- | --- |
| Thailand | Producers pay for RWM | Governmental funding | Producer's responsibility |
| Türkiye | Producers pay for RWM | Producers have to pay fee to Radioactive Waste Management Account | Operators have to pay fee to Decommissioning Account |
| Ukraine | Producers pay for RWM, National RW Management Fund | National RW Management Fund | Decommissioning Fund, governmental funding for Chornobyl NPP |
| United Arab Emirates | Producers pay for RWM | Fund for NPP operators — Decommissioning Trust Fund | Fund for NPP operators — Decommissioning Trust Fund |
| United Kingdom | Producers pay for RWM | Producer's responsibility, governmental funding, Nuclear Liabilities Fund | Governmental funding for decommissioning costs for the NDA estate. Decommissioning costs for the currently existing AGR and PWR reactors will be met through the Nuclear Liabilities Fund |
| United States of America | Producers pay for RWM | Private operators of facilities need to demonstrate capability to fund operational waste management. Public facilities obtain government funding. For spent fuel and HLW disposal operators have paid into a Nuclear Waste Fund (currently suspended) | For NPPs have decommissioning funds, in case of non-legacy sites producer pays |
| Uruguay | Producers pay for RWM | n.a. | n.a. |
| Uzbekistan | Producers pay for RWM | Governmental funding | Governmental funding |
| Viet Nam | Producers pay for RWM | Producers contribute to National Radioactive Waste. Management Fund to be established | Decommissioning Fund from fees on nuclear energy |

# ABBREVIATIONS

| | |
|---|---|
| DSRS | disused sealed radioactive source |
| Euratom | European Atomic Energy Community |
| EC | European Commission |
| EW | exempt waste |
| HLW | high level waste |
| ILW | intermediate level waste |
| LLW | low level waste |
| MOX | mixed oxide |
| NORM | naturally occurring radioactive material |
| NPP | nuclear power plant |
| NEA | Nuclear Energy Agency |
| SRIS | Spent Fuel and Radioactive Waste Information System |
| t HM | tonnes of heavy metal |
| UMM | uranium mining and milling |
| VLLW | very low level waste |
| VSLW | very short lived waste |
| WMO | waste management organization |
| WNA | World Nuclear Association |

# CONTRIBUTORS TO DRAFTING AND REVIEW

| | |
|---|---|
| Erim, A. | World Nuclear Association |
| Evans, C. | Orano, France |
| Garcia Neri, E. | Enresa, Spain |
| Gastl, C. | International Atomic Energy Agency |
| Gonzalez-Espartero, A. | International Atomic Energy Agency |
| Hedberg, B. | Swedish Radiation Safety Authority, Sweden |
| Lazar, I. | Hungarian Atomic Energy Authority, Hungary |
| Lust, M. | International Atomic Energy Agency |
| Matuzas, V. | European Commission |
| Robbins, R. | International Atomic Energy Agency |
| Shenk, J. | Department of Energy, United States of America |
| Suárez Alvarado, R. | National Nuclear Safety and Safeguards Commission, Mexico |
| Svedkauskaite, J. | European Commission |
| Ünver, L.O. | Turkish Atomic Energy Authority, Türkiye |
| Yilma, H. | Nuclear Energy Agency |

**Technical Meetings**

Vienna, Austria: 21–25 February 2022,
14–16 February 2023

**Consultants Meetings**

Vienna, Austria: 20–23 October 2020, 12 January–30 June 2021,
1 September–29 October 2021, 23–25 November 2021,
22–25 February 2022, 1–3 November 2022,
14–16 February 2023

# Structure of the IAEA Nuclear Energy Series*

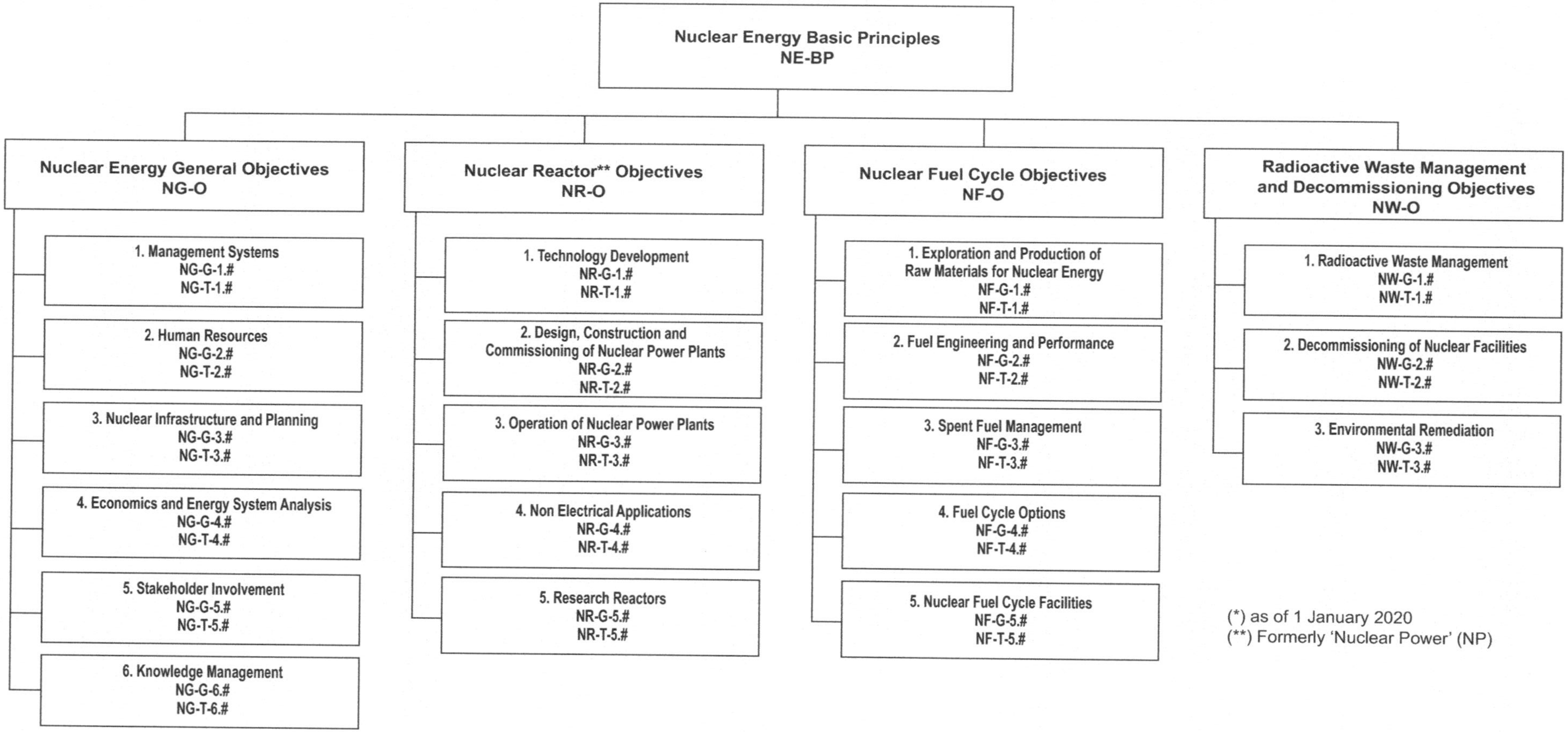

**Nuclear Energy Basic Principles
NE-BP**

**Nuclear Energy General Objectives
NG-O**

1. Management Systems
NG-G-1.#
NG-T-1.#

2. Human Resources
NG-G-2.#
NG-T-2.#

3. Nuclear Infrastructure and Planning
NG-G-3.#
NG-T-3.#

4. Economics and Energy System Analysis
NG-G-4.#
NG-T-4.#

5. Stakeholder Involvement
NG-G-5.#
NG-T-5.#

6. Knowledge Management
NG-G-6.#
NG-T-6.#

**Nuclear Reactor** Objectives
NR-O**

1. Technology Development
NR-G-1.#
NR-T-1.#

2. Design, Construction and Commissioning of Nuclear Power Plants
NR-G-2.#
NR-T-2.#

3. Operation of Nuclear Power Plants
NR-G-3.#
NR-T-3.#

4. Non Electrical Applications
NR-G-4.#
NR-T-4.#

5. Research Reactors
NR-G-5.#
NR-T-5.#

**Nuclear Fuel Cycle Objectives
NF-O**

1. Exploration and Production of Raw Materials for Nuclear Energy
NF-G-1.#
NF-T-1.#

2. Fuel Engineering and Performance
NF-G-2.#
NF-T-2.#

3. Spent Fuel Management
NF-G-3.#
NF-T-3.#

4. Fuel Cycle Options
NF-G-4.#
NF-T-4.#

5. Nuclear Fuel Cycle Facilities
NF-G-5.#
NF-T-5.#

**Radioactive Waste Management and Decommissioning Objectives
NW-O**

1. Radioactive Waste Management
NW-G-1.#
NW-T-1.#

2. Decommissioning of Nuclear Facilities
NW-G-2.#
NW-T-2.#

3. Environmental Remediation
NW-G-3.#
NW-T-3.#

(*) as of 1 January 2020
(**) Formerly 'Nuclear Power' (NP)

*Key*
**BP:** Basic Principles
**O:** Objectives
**G:** Guides and Methodologies
**T:** Technical Reports
**Nos 1–6:** Topic designations
**#:** Guide or Report number

*Examples*
**NG-G-3.1:** Nuclear Energy General (**NG**), Guides and Methodologies (**G**), Nuclear Infrastructure and Planning (topic **3**), **#1**
**NR-T-5.4:** Nuclear Reactors (**NR**), Technical Report (**T**), Research Reactors (topic **5**), **#4**
**NF-T-3.6:** Nuclear Fuel (**NF**), Technical Report (**T**), Spent Fuel Management (topic **3**), **#6**
**NW-G-1.1:** Radioactive Waste Management and Decommissioning (**NW**), Guides and Methodologies (**G**), Radioactive Waste Management (topic **1**) **#1**

# CONTACT IAEA PUBLISHING

Feedback on IAEA publications may be given via the on-line form available at:
www.iaea.org/publications/feedback

This form may also be used to report safety issues or environmental queries concerning IAEA publications.

Alternatively, contact IAEA Publishing:

Publishing Section
International Atomic Energy Agency
Vienna International Centre, PO Box 100, 1400 Vienna, Austria
Telephone: +43 1 2600 22529 or 22530
Email: sales.publications@iaea.org
www.iaea.org/publications

Priced and unpriced IAEA publications may be ordered directly from the IAEA.

## ORDERING LOCALLY

Priced IAEA publications may be purchased from regional distributors and from major local booksellers.

Printed and bound by CPI Group (UK) Ltd, Croydon, CR0 4YY

06/07/2026

02160600-0010